# Women's Special Issue Series:
Analytical Methods

# Women's Special Issue Series: Analytical Methods

Editors

**Victoria Samanidou**
**Verónica Pino**
**Natasa Kalogiouri**

Basel • Beijing • Wuhan • Barcelona • Belgrade • Novi Sad • Cluj • Manchester

*Editors*

Victoria Samanidou
Chemistry- Laboratory of
Analytical Chemistry
University of Thessaloniki
Thessaloniki
Greece

Verónica Pino
Chemistry
Universidad de La Laguna
La Laguna
Spain

Natasa Kalogiouri
Chemistry- Laboratory of
Analytical Chemistry
University of Thessaloniki
Thessaloniki
Greece

*Editorial Office*
MDPI
St. Alban-Anlage 66
4052 Basel, Switzerland

This is a reprint of articles from the Special Issue published online in the open access journal *Methods and Protocols* (ISSN 2409-9279) (available at: www.mdpi.com/journal/mps/special_issues/women_SI_Analytical_Methods).

For citation purposes, cite each article independently as indicated on the article page online and as indicated below:

Lastname, A.A.; Lastname, B.B. Article Title. *Journal Name* **Year**, *Volume Number*, Page Range.

**ISBN 978-3-0365-9147-6 (Hbk)**
**ISBN 978-3-0365-9146-9 (PDF)**
doi.org/10.3390/books978-3-0365-9146-9

© 2023 by the authors. Articles in this book are Open Access and distributed under the Creative Commons Attribution (CC BY) license. The book as a whole is distributed by MDPI under the terms and conditions of the Creative Commons Attribution-NonCommercial-NoDerivs (CC BY-NC-ND) license.

# Contents

About the Editors . . . . . . . . . . . . . . . . . . . . . . . . . . . . . . . . . . . . . . . . . . . . . . . . . . . . . . . . vii

Preface . . . . . . . . . . . . . . . . . . . . . . . . . . . . . . . . . . . . . . . . . . . . . . . . . . . . . . . . . . . . . . ix

**Juan Aspromonte, Carlina Lancioni and Giorgia Purcaro**
Solid-Phase Microextraction—Gas Chromatography Analytical Strategies for Pesticide Analysis
Reprinted from: *Methods Protoc.* 2022, 5, 82, doi:10.3390/mps5050082 . . . . . . . . . . . . . . . . . 1

**Vasiliki Keramari, Sophia Karastogianni and Stella Girousi**
New Prospects in the Electroanalysis of Heavy Metal Ions (Cd, Pb, Zn, Cu): Development and Application of Novel Electrode Surfaces
Reprinted from: *Methods Protoc.* 2023, 6, 60, doi:10.3390/mps6040060 . . . . . . . . . . . . . . . . . 18

**Despoina Giamaki, Konstantina Dindini, Victoria F. Samanidou and Maria Touraki**
Simultaneous Quantification of Bisphenol-A and 4-Tert-Octylphenol in the Live Aquaculture Feed *Artemia franciscana* and in Its Culture Medium Using HPLC-DAD
Reprinted from: *Methods Protoc.* 2022, 5, 38, doi:10.3390/mps5030038 . . . . . . . . . . . . . . . . . 33

**Ana Castiñeira-Landeira, Lua Vazquez, Thierry Dagnac, Maria Celeiro and María Llompart**
Allergens and Other Harmful Substances in Hydroalcoholic Gels: Compliance with Current Regulation
Reprinted from: *Methods Protoc.* 2023, 6, 95, doi:10.3390/mps6050095 . . . . . . . . . . . . . . . . . 49

**Leyla Karadurmus, Sevinc Kurbanoglu, Bengi Uslu and Sibel A. Ozkan**
An Efficient, Simultaneous Electrochemical Assay of Rosuvastatin and Ezetimibe from Human Urine and Serum Samples
Reprinted from: *Methods Protoc.* 2022, 5, 90, doi:10.3390/mps5060090 . . . . . . . . . . . . . . . . . 71

**Cemile Yücel, Ilgi Karapinar, Serenay Ceren Tüzün, Hasan Ertaş and Fatma Nil Ertaş**
Determination of UV Filters in Waste Sludge Using QuEChERS Method Followed by In-Port Derivatization Coupled with GC–MS/MS
Reprinted from: *Methods Protoc.* 2022, 5, 92, doi:10.3390/mps5060092 . . . . . . . . . . . . . . . . . 81

**Sergi Mallorca-Cebria, Yolanda Moliner-Martinez, Carmen Molins-Legua and Pilar Campins-Falcó**
On-Site Multisample Determination of Chlorogenic Acid in Green Coffee by Chemiluminiscent Imaging
Reprinted from: *Methods Protoc.* 2023, 6, 20, doi:10.3390/mps6010020 . . . . . . . . . . . . . . . . . 95

**Thasmin Shahjahan, Bilal Javed, Vinayak Sharma and Furong Tian**
pH and NaCl Optimisation to Improve the Stability of Gold and Silver Nanoparticles' Anti-Zearalenone Antibody Conjugates for Immunochromatographic Assay
Reprinted from: *Methods Protoc.* 2023, 6, 93, doi:10.3390/mps6050093 . . . . . . . . . . . . . . . . . 105

**Annisa Nurkhasanah, Titouan Fardad, Ceferino Carrera, Widiastuti Setyaningsih and Miguel Palma**
Ultrasound-Assisted Anthocyanins Extraction from Pigmented Corn: Optimization Using Response Surface Methodology
Reprinted from: *Methods Protoc.* 2023, 6, 69, doi:10.3390/mps6040069 . . . . . . . . . . . . . . . . . 121

**Fausto Viteri, Nazly E. Sánchez and Katiuska Alexandrino**
Determination of Polycyclic Aromatic Hydrocarbons (PAHs) in Leaf and Bark Samples of *Sambucus nigra* Using High-Performance Liquid Chromatography (HPLC)
Reprinted from: *Methods Protoc.* 2023, 6, 17, doi:10.3390/mps6010017 . . . . . . . . . . . . . . . . . 134

# About the Editors

**Victoria Samanidou**

Dr Victoria Samanidou is Full Professor, Director of the Laboratory of Analytical Chemistry, and Vice-President of the School of Chemistry of Aristotle University of Thessaloniki, Greece. Her research interests focus on the development of sample preparation methods using sorptive extraction prior to chromatographic analysis. She has co-authored 218 original research articles in peer-reviewed journals and 67 reviews, 85 editorials/in view/opinions/commentaries, and 57 chapters in scientific books (h-index 44, 6748 citations). She is an Editorial Board Member of more than 31 scientific journals and has been the Guest Editor of more than 32 Special Issues. She has peer reviewed more than 760 manuscripts for 161 scientific journals. Her educational activities focus on instrumental chemical analysis and separation science. She has participated in numerous scientific conferences (also as member of organizing and/or scientific committee) and she has organized several scientific events. In 2016, she was included in top 50 Power List of women in analytical science, as proposed by Texere Publishers, while in 2021 she was included in the "The Analytical Scientist" 2021 Power List of top 100 influential people in analytical science. In 2023, she was included in the Power List 2023 in the following category: Mentors and Educators. Since 2020, she has been included in the top 2% of world scientists in the field of analytical chemistry (career and single year) published in PLOS Biology based on citations from SCOPUS. She is also enlisted as one of the 50 scientists from the Aristotle University of Thessaloniki Scientist in AD University Rankings 2023.

**Verónica Pino**

Dr. Verónica Pino is Full Professor at the University of La Laguna (ULL), Spain, and member of the Institute of Tropical Diseases and Public Health of ULL. Her research activities intend to showcase advances in analytical sample preparation and novel microextraction materials, with high emphasis on sustainability. She has published more than 120 publications in high-impact journals indexed by the *Journal of Citation Reports*, with an h index of 44 (Google Scholar), mostly derived from research articles and not from review articles, thus pointing out the impact of her research. She is the corresponding authors in most of these publications. In addition, she is author of more than 25 book chapters belonging to top chemistry editorials, and Editor of a series of 5 books for Wiley (*Analytical Sample Preparation*) and 1 ongoing book for Elsevier (*Metal-Organic Frameworks in Analytical Sample Preparation and Sensing*). She is currently an Editorial Board Member of the prestigious journal *Analytica Chimica Acta* (first quartile in Analytical Chemistry). She has directed seven PhD theses, and another five PhD theses are under development in her group. Dr. Pino has presented more than a hundred scientific communications in prestigious international congresses, with more than forty invited keynotes and oral communications, and one plenary lecture. She has participated in the organization of several scientific events and was the Chairwoman of the International Conference ExTech 2023, held in Tenerife in July 2023. Dr. Pino is the Principal Investigator of the MAT4LL group, with one consecutive projects as principal investigator with funding from the highly competitive calls of the Spanish Agency of Research, and was more recently awarded with one proof-of-concept project related to innovation and transference, and two patents, one of them already licensed to a company.

**Natasa Kalogiouri**

Dr Natasa Kalogiouri is currently an Assistant Professor of Analytical Chemistry in the Laboratory of Analytical Chemistry, Department of Chemistry, Aristotle University of Thessaloniki (AUTH). Her research interests focus on the development of analytical methods using microextraction techniques prior to liquid (HPLC-UV/DAD, LC-MS, LC-MS/MS, LC-QTOF-MS/MS) and/or gas chromatographic analysis (GC-FID, GC-MS, GC×GC-MS), and the use of targeted and non-targeted screening workflows in combination with advanced chemometric techniques. She has developed in-house chemometric tools for data mining (data analysis, modeling, classification, and forecasting). She has worked on 15 scientific programs and counts more than 55 scientific papers in peer-reviewed journals with an h-index of 15 (Google Scholar), 50 conference announcements, and 4 book chapters. She has served as Guest Editor of four Special Issues. She has been a member of the organization/scientific committee in 12 International and National Conferences. She is a member of the Association of Greek Chemists and Member of European Chemical Society—Division of Analytical Chemistry (EuChemS-DAC).

# Preface

Undoubtedly, all pivotal advances in a great number of scientific fields rely on the respective advances in chemistry. Among all fields, analytical chemistry has the leading role. Analytical methods are necessary in dentistry, medicine (both human and veterinary), archaeology, the pharmaceutical industry, food science and technology, and environmental sciences, and these are only a few examples. The multidisciplinary role of chemistry is reflected in all important advances from research groups in every technological progress, proving that analytical chemistry is the key issue in scientific progress.

In parallel with this outstanding role of analytical chemistry in a variety of scientific fields, we would particularly like to reflect the impact of female researchers in the field of analytical chemistry in this Special Issue to serve as a motivation guide for girls and women pursuing a STEM career.

Therefore, we have invited well-established scientists to share the results of their research with the scientific community through this Special Issue, which aimed to compile manuscripts written or lead by women analytical chemists.

In this Special Issue, seven original research articles, two review articles, and one protocol show the impact of female researchers in the field of analytical science.

The Guest Editors would like to thank all authors for their fine contributions.

**Victoria Samanidou, Verónica Pino, and Natasa Kalogiouri**
*Editors*

Review

# Solid-Phase Microextraction—Gas Chromatography Analytical Strategies for Pesticide Analysis

Juan Aspromonte [1], Carlina Lancioni [1] and Giorgia Purcaro [2,*]

1 Laboratorio de Investigación y Desarrollo de Métodos Analíticos, LIDMA, Facultad de Ciencias Exactas (Universidad Nacional de La Plata, CIC-PBA, CONICET), Calle 47 esq. 115, La Plata 1900, Argentina
2 Gembloux Agro-Bio Tech, University of Liège, Passage des Déportés 2, 5030 Gembloux, Belgium
* Correspondence: gpurcaro@uliege.be; Tel.: +32-(0)81-62-22-20

**Abstract:** Due to their extensive use and the globalized commerce of agricultural goods, pesticides have become a global concern. Despite the undoubtful advantages of their use in agricultural practices, their misuse is a threat to the environment and human health. Their analysis in environmental samples and in food products continues to gain interest in the analytical chemistry community as they are challenging matrices, and legal concentration limits are particularly low (in the order of ppb). In particular, the use of solid-phase microextraction (SPME) has gained special attention in this field thanks to its potential to minimize the matrix effect, while enriching its concentration, allowing very low limits of detection, and without the need of a large amount of solvents or lengthy procedures. Moreover, its combination with gas chromatography (GC) can be easily automated, making it a very interesting approach for routine analysis. In this review, advances and analytical strategies for the use of SPME coupled with GC are discussed and compared for the analysis of pesticides in food and environmental samples, hopefully encouraging its further development and routine application in this field.

**Keywords:** SPME; pesticide; gas chromatography; mass spectrometry; food; environmental

## 1. Introduction

Since its introduction in the early 1990s [1], solid-phase microextraction (SPME) methodologies have been widely explored, proving that it constitutes a non-exhaustive analytical tool which is versatile, appropriate for simple, and effective as sample pretreatment and/or preconcentration step for the analysis of a broad range of analytes in plenty of studies distributed over a wide variety of areas, such as flavors and fragrances, metabolomics, pharmaceutical, and biomedical analysis [2–6]. As an environmentally friendly sampling technology requiring a minimum or zero amount of solvents, it is not surprising that a large amount of papers are devoted to the use of SPME in the analysis of common environmental contaminants, such as polychlorinated biphenyls, polycyclic aromatic hydrocarbons, and pesticides, etc. [7–9].

Pesticides are defined as a substance or mixture of substances intended for preventing, destroying, repelling or mitigating any pest; use as plant regulator, defoliant or desiccant or as nitrogen stabilizer [10]. These numerous groups of substances might be categorized according to their origin in natural or synthetic; to their biocide function (i.e., the target pest object) or more frequently, according to their particular moieties in four main types: Organochlorine pesticides (OCPs), organophosphorus pesticides (OPPs), organonitrogen pesticides (ONPs), and pyrethroids [11]. The use of pesticides, mainly in agriculture, is a usual practice that has been spreading worldwide since the mid-20th century to increase crop productivity and preservation, aiming to fulfill the alimentary supply demanded by a continuously growing population [12,13]. Despite the undoubted social and economic advantages of this practice, the vast application of pesticides might produce adverse effects, which can be worsened if good agricultural practices are not satisfied, such as improper handling during the application of these products. On the one hand, a risk of environmental contamination exists. Typically, soil acts as the main pesticide receptor. Indeed, pesticides reach the ground due to a direct soil application or, indirectly, after an application on crops. Once they reach the soil, they will persist and also dissipate toward water bodies and air through wind and rain, leaching, runoff or volatilization. On the other hand, pesticides possess a certain chemical stability and ability to bioaccumulate. Therefore, they represent an important hazard to human health. They may reach the human body by ingesting contaminated drinking water and food (fruit, vegetable, fish, honey, milk, etc.), inhalation of contaminated air or dermal contact with pesticide-contaminated water, air or soil [14].

Considering that this is an issue of global concern, regulatory bodies of multiple jurisdictions have made decisions in this regard, promulgating limit values for pesticides in soil, air, drinking water, and agricultural commodities. The limit values vary in several orders of magnitude depending on the country, pesticide, and matrix. However, they often are as low as parts-per-billon, with some even lower limits [15]. In this sense, the need for accurate, sensitive, and robust analytical methods for pesticide determination at trace levels in a wide diversity of matrices becomes evident. Prior to the instrumental analysis, which is usually a chromatographic method, sample preparation procedures have to be applied to isolate analytes from their matrices, remove interferences, and increase the concentrations to detectable values.

Reported procedures for pesticide analysis usually include the use of liquid-liquid extraction (LLE), solid-phase extraction (SPE), accelerated solvent extraction, and quick, easy, low-cost, effective, rugged, and safe methods (QuEChERS). However, these sample pretreatments are very time-consuming, involve procedures with multiple stages, and present relatively high solvent consumption, which is an attempt against the green chemistry principle of reducing waste. Moreover, these techniques often yield poor analyte enrichment or selectivity, therefore requiring the use of sophisticated analytical instruments to compensate for the lack of selectivity in the sample preparation, such as tandem mass spectrometry (MS/MS), which will require additional maintenance and is only exploited to deal with the matrix.

In this regard, SPME has gained special attention in pesticide determination thanks to its capacity to overcome the abovementioned limitations. Furthermore, compared to liquid chromatography, if SPME is combined with gas chromatography (GC) analysis and thermal desorption as the introduction mode into the instrument, most of the principles of green analytical chemistry are fulfilled [16,17]. Moreover, automatic systems are readily available, increasing their application potential. Therefore, in this review, we focus exclusively on GC. Even though it could be argued that liquid chromatography is the general method of choice for the determination of polar and less volatile compounds, it could be attributed to the compatibility of extracts obtained by traditional preconcentration procedures. At present, with the advances in the development of GC columns and the possibility of tuning analyte chemical properties by derivatization procedures (which can be performed onto the coating), this does not represent an inconvenience if automation and greenness are pondered.

When the Scopus bibliometric database [18] is employed to carry out a systematic search within the article title, abstract, and keywords, including the combined descriptors "pesticides", "gas chromatography", and "SPME", the number of published documents by year can be depicted as shown in Figure 1. As can be seen, the interest in using this technique has constantly been growing in this field of application, and it is expected to continue in this direction. A posterior selection of articles to be included in this review was carried out, maintaining those which address the optimization of SPME conditions to some extent.

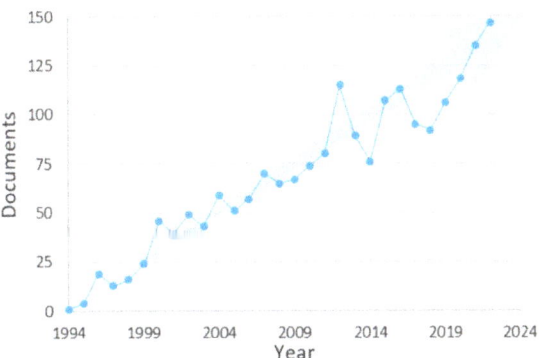

**Figure 1.** The number of published articles per year from the Scopus bibliometric database [10] using "pesticides", "gas chromatography", and "SPME" as descriptors. The number of publications by the end of the current year is extrapolated from the existing data.

Despite the numerous reviews on SPME in many fields of applications [19,20], none focus on the advantages of the use of SPME in the cumbersome and highly regulated area of pesticides. In this review, advances in the analysis of pesticides in environmental and food samples using SPME coupled to GC are discussed and compared. Although essential for the betterment of agricultural practices and environmental protection, articles dealing with straightforward applications of the technique without further optimization of the analytical techniques involved, are not considered here. This review is not intended as an exhaustive literature compilation of all the work carried out in this area and characteristic cases were selected to exemplify the different advances in the field. Moreover, the principles of the technique are out of the aim of this review and this information has been largely discussed recently elsewhere [21,22].

## 2. General Considerations

The underlying principle of the SPME methodology is based on the partitioning of analytes between an extractant phase and a sample (liquid or solid). To date, in the fiber approach, the most used geometry which is the extractant phase or sorbent (polymeric liquid, solid sorbent or a combination of both) is deposited on a thin fused silica or metallic rod, which acts as support. Once the fiber is exposed to the gas phase above the sample (HS-SPME) or directly into the sample (DI-SPME), the mass transfer takes place driven by the concentration gradient generated among the two or three phases, respectively. Once the microextraction is performed, desorption is conducted by immersion of the fiber into a compatible solvent or thermally in the injection port of a gas chromatograph.

In SPME method development, it is essential to count with information about the physicochemical properties of both sample matrix and analyte, to set adequate extraction conditions for the attainment of the desired performance of the procedure. Extraction conditions to be established include several variables, such as operation mode, extraction temperature, exposure time, volume phase ratio, pH, ionic strength and pressure, desorption mode, etc. Nevertheless, due to the large number of parameters to be optimized for the SPME and since they could be interdependent, the use of the design of experiments (DoE) should be considered. Combining prior information with simple exploratory designs (Plackett-Burman, Pareto, etc.), the number of parameters that require optimization can be reduced. Then, more complete designs, such as response surfaces, can be used for the remaining ones. However, a key aspect of the overall SPME process relies on sorbent selection. The available marketed sorbents cover a wide range of polarity, volatility, and molecular weight, which are commercialized fibers with single-phase liquid polymeric materials, such as polydimethylsiloxane (PDMS), poly(oxi)ethyleneglycol (PEG), and polyacrylate (PA), as well as mixed-phase materials consisting of solid carboxen (CAR) and/or divinyl benzene (DVB) particles dispersed in PDMS [20,23]. Additionally, research on novel materials development is constantly taking place to improve selectivity and extraction efficiency, reduce costs, increase reusability, enhance porosity and surface area, and provide better chemical, mechanical, and/or thermal stability, etc. Emerging materials employed as SPME coatings include molecularly imprinted polymers (MIPs), metal organic frameworks (MOFs), ionic liquid (IL)-based sorbents, carbon-based materials (graphene, fullerenes, and carbon nanotubes), conducting polymers, monolith-based sorbents and composites [22–24]. Concerning pesticide analysis, it can be seen in the following sections that applications involve both commercially available sorbents and novel materials developed for a specific group of pesticides and also to perform multiresidue analysis in different matrices. Details on specific reported works employing novel coatings will be further discussed in following sections.

In addition to the development of new coating materials, there have been advances to increase the sorbent capacity (larger sorbent volumes), while improving the lifespan of the SPME fiber. This has been achieved by changing the geometry of classical fibers into the so-called "Arrow" and HiSorb™ SPME. In the first case, an arrow-shaped metal rod is coated with a sorbent material (similar to those used for classic SPME), obtaining a sorbent volume up to 11.8 µL; this is more than 13 times the volume of a classic 100 µm PDMS fiber (~0.9 µL) [25]. This larger sorbent volume can increase recoveries and improve the extraction capacity of polar compounds, achieving better sensitivity [26]. Moreover, the arrow shape confides a better mechanical stability [27]. In the case of HiSorb™, the idea of increasing sensitivity through a larger sorbent volume has been expanded to obtain 63 µL of sorbent [25]. This came at the cost of considerably enlarging the size of the metal rod, to the point of requiring dedicated modules for desorption, jeopardizing the straightforward compatibility of SPME with GC. Nevertheless, a fully automated platform (Centri) has been developed in 2018 by Markes Int. to overcome this issue [25]. Despite the great potential of these SPME variants, they have not yet been applied to the analysis of pesticides to the best of our knowledge.

Another interesting alternative to increase the sorbent capacity is the use of thin films [28]. These devices are somehow similar to a classic SPME fiber, except that the extraction phase is deposited on a flat substrate, increasing its surface. Its application to the determination of pesticides in water showed promising results for a lab-made device, reaching limits of detection in the range of 20 to 300 ppt when in combination with GC-MS [29]. Nevertheless, the main drawback of these devices is that they are not compatible with GC injector ports, thus they require a compatible thermal desorption system, making their automation less straightforward than the SPME fiber.

Regarding the chromatographic method for pesticide analysis by SPME, there are several reasons to affirm that GC represents a better alternative compared to liquid chromatography. GC possesses better attributes in terms of solvent consumption (greener approach), automation, and cost-efficiency. Limitations in the use of GC for very polar and thermally labile pesticides can be overcome by derivatization procedures and an appropriate column selection [30]. Derivatization can be carried out prior to SPME extraction by adding a derivatizing agent directly into the sample matrix. Then, the derivatized product can be extracted by SPME. Apart from its simplicity, this approach has the advantage of potentially increasing the analyte recovery. For instance, the use of this approach in combination with MS detection allowed limits of detection as low as 2 pg m$^{-3}$ for the determination of pesticides in air [31]. However, possible interferences from the matrix during derivatization should be considered [32]. An alternative to the derivatization prior to SPME, is to use the injection port of the GC to derivatize the analytes after SPME extraction. In this case, a derivatization agent is injected into the GC injection port only prior to the desorption of the SPME fiber. The application of this technique in combination with GC-MS/MS allowed for the attainment of limits of detection in the range of 0.04 to 0.24 ng m$^{-3}$ for the determination of eight pesticides in air. Of note, in these cases, SPME acts as a preconcentration step prior to GC analysis, as the analytes are first collected in air samplers. A third type of derivatization strategy consists of loading a derivatizing agent into the SPME fiber prior to extraction, performing a simultaneous extraction and derivatization on the fiber. This is certainly advantageous in terms of manipulations; however, it may not be simple to find an appropriate derivatizing reagent and the optimization of the process may be cumbersome [33]. Nevertheless, it should be considered that the use of derivatization techniques should be minimized to the greatest possible extent to not hinder the green analytical chemistry potential of SPME [34].

Another important aspect to consider is the detection system to which GC is coupled. Herein, selective detectors exist, which offer good sensitivity and enough specificity toward certain classes of pesticides. For instance, electron capture detection (ECD) is widely used for OCPs, while nitrogen-phosphorus detection (NPD), responsive to compounds containing N and P atoms, is recommended for ONPs and OPPs, respectively. Similarly, the flame photometric detector (FPD), which is sensitive to sulphur- or phosphorous-containing compounds, is also recommended for OPPs analysis. The main drawback of these selective detectors is that their applicability is limited to specific pesticide classes and usually selectivity is not sufficient since there are numerous interfering compounds related to the pesticides of interest, such as transformation products or metabolites. In this sense, MS detection arises as a solution for selectivity and specificity issues. Of note, MS constitutes an extra separation dimension that also allows for the identification of the compounds being separated. However, the use of the full scan mode, rather than selective ion monitoring (SIM) during MS acquisition, guarantee a reliable identification of the analytes, allowing for the performance of untargeted and post-targeted approaches. Nevertheless, the complexity of the samples can hinder this capability and, in particular, can be detrimental for a reliable quantification. For this reason, some authors proposed the use of two stages, first an identification stage using MS in full scan, although the separation of the compounds may not be good for quantification, and then a quantification stage using MS or MS/MS with selected ions for the pesticides that could be identified [35–37]. Despite the complexity of the chromatogram obtained for these samples, the use of MS, and, in particular, MS/MS facilitates the quantification of target analytes (predefined or identified during a screening run), enabling multiresidues determinations without a full chromatographic separation (e.g., Figure 2) [35,38–40]. When selected ions are used, the identity of the peak can be confirmed by monitoring at least three ions when operating the MS in selective ion monitoring (SIM) or two or three transitions in the case of multiple reaction monitoring (MRM) for MS/MS. For instance, Al-Alam et al. [38] developed a method for the determination of 60 pesticides in honey with limits of detection ranging from 0.12 to 50.42 ppb by means of MS/MS monitoring one precursor ion and one or two product ions for the selected analytes as confirmatory ions.

An interesting alternative to unriddle the complex chromatograms that are obtained for real samples is the use of multidimensional chromatography, which can enable more reliable identification of pesticides through better chromatographic separation. For instance, the use of 2D heart-cut GC allowed for a better resolution of poorly resolved small analyte peaks that overlap with large matrix ones [41,42]. This allowed for identification with a single quadrupole MS of the analytes of interest and it could be used for quantification of the targeted analytes to the sub-ppt levels. Surprisingly, to the best of our knowledge, the combination of the extraction and enrichment capacity of SPME and the use of comprehensive multidimensional gas chromatography (GC×GC) applied to pesticides analysis has been reported only for drinking water [43]. This technique can provide the needed separation power to obtain clearly separated peaks for better identification, facilitating untargeted pesticide screening analysis in challenging samples, such as food and soil, while combining the quantitative enhancement and automation possibilities of SPME.

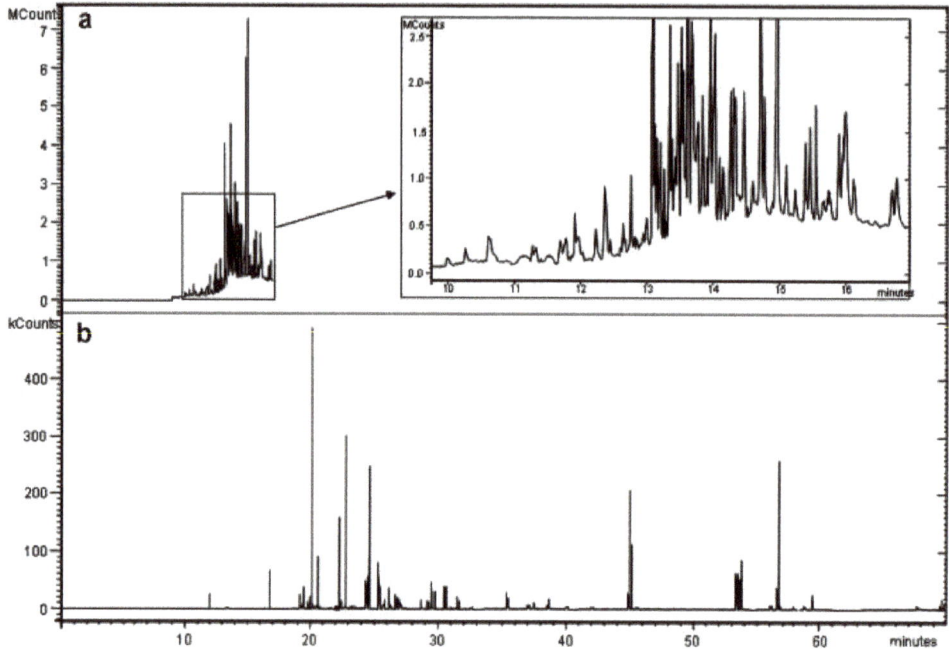

**Figure 2.** Chromatograms obtained for the analysis by the SPME-GC method of an orange juice spiked with multiple target pesticides at a concentration of 0.05 mg L$^{-1}$ (**a**) obtained with the GC-MS screening full scan method and (**b**) obtained with the confirming/quantifying MS/MS method. Reprinted with permission from Ref. [39]; 2008, Elsevier.

**Table 1.** Summary of the methods discussed in this review for the two main groups of samples, food and environmental.

| Matrix | SPME Coating Tested [1] | SPME Mode | GC Detector | LOD | | Ref. |
|---|---|---|---|---|---|---|
| | Food Samples | | | | | |
| Fruits and Vegetables | PA PDMS PDMS/DVB PDMS/CAR DVB/CAR/PDMS | DI | MS (SIM) | 1.00–10.00 | ppb | [36] |
| | C-(C3N4@MOF) | DI | MS (SIM) | 0.23–7.5 | ppb | [44] |
| | PDMS/DVB | DI | MS (TIC) [2] | 0.013–0.110 (for 2D) | ppt | [42] |
| | PA | DI | MS (identification) FPD (quantification) | 0.01–0.14 | ppb | [45] |
| | PDMS (modified) | DI | TOFMS | 1–50 | ppb | [46] |
| | COF | DI | ECD | 0.04–0.25 | ppb | [47] |
| | PDMS/DVB/PDMS | DI | MS (SIM) | 1.0–33.0 only LOQ reported | ppb | [48] |
| | IL on silica | HS | FID | 0.01–0.93 | ppb | [49] |
| | COF | HS | ECD | 0.0003–0.0023 | ppt | [50] |
| | PDMS | HS | ECD | 0.01–1.0 | ppb | [51] |
| | PDMS PDMS/DVB | HS | MS (TIC) MS (SIM) | 0.11–3.48 | ppb | [35] |

Table 1. Cont.

| Matrix | SPME Coating Tested [1] | SPME Mode | GC Detector | LOD | | Ref. |
|---|---|---|---|---|---|---|
| Wine and Juice | PA | DI | MS (SIM)<br>NPD<br>FID | 0.01–15<br>10–6000<br>200–19000 | ppt | [52] |
| | PA | DI | MS (SIM) | 2–90 | ppb | [53] |
| | PDMS<br>PDMS/DVB | DI | MS/MS | 0.8–19.6 | ppb | [39] |
| | PDMS/DVB | HS | MS (TIC) [2] | 0.062–33.515 (for 2D) | ppb | [41] |
| Milk | PDMS/DVB | DI | µECD | 0.003–0.56 | ppb | [54] |
| | PDMS<br>PDMS/DVB | DI<br>HS | MS/MS | 0.01–0.30 | ppb | [40] |
| | PDMS/DVB | HS | MS (SIM) | 2.2–10.9 | ppb | [37] |
| Honey | PA<br>PDMS | DI | MS/MS | 0.12–50.42 | ppb | [55] |
| | PDMS<br>PA | DI | AED | 0.02–10.0 | ppb | [56] |
| | Electrospun nanostructured PS | HS | MS (SIM) | 0.1–2 | ppb | [57] |
| Environmental samples | | | | | | |
| Soil and sediment | PDMS | DI | MS (identification)<br>ECD (quantification) | 0.6–30 | ppb | [58] |
| | PA | DI | MS (TIC) | 0.1–60 | ppb | [59] |
| | PA | DI | MS (SIM)<br>NPD<br>FID | 0.01–15<br>10–6000<br>200–19000 | ppt | [52] |
| Water (including drinking water) | PDMS/DVB | DI | MS (TIC)<br>ECD<br>NPD | 4–32 | ppt | [60] |
| | PA | DI | MS (SIM) | 0.05–19 | ppb | [61] |
| | PA | DI | MS (SIM) | 3–200 | ppt | [62] |
| | PDMS/DVB | DI | MS (SIM) | 0.003–0.322 | ppb | [63] |
| | PDMS/DVB | DI | MS/MS | 0.0002–0.04 | ppb | [64] |
| | NU-1000 (MOF) | DI | MS (SIM) | 0.011–0.058 | ppt | [65] |
| | DVB/CAR/PDMS | DI | ECD | 0.001–0.45 | ppt | [66] |
| | DVB/CAR/PDMS | DI | ECD | 0.002–0.070 | ppb | [67] |
| | PDMS/DVB | DI | ECD | 2.6–5.7 | ppt | [68] |
| | PDMS | DI | MS [3] | 0.001–0.025 | ppb | [43] |
| | Nafion on SBA-15 | HS | MS (TIC) | 0.01–0.09 | ppb | [69] |
| | PDMS<br>PDMS/DVB | HS | MS/MS | 0.9–26.3 | ppt | [70] |
| | PA | HS | HRMS (magnetic sector) | 0.01–350 only LOQ reported | ppt | [71] |
| | PDMS | HS | ECD | 0.034–0.301 | ppb | [72] |

[1] PA: Polyacrylate; PDMS: Polydimethylsiloxane; DVB: Divinylbenzene; CAR: Carboxen; MOF: Metal organic framework; COF: Covalent organic framework; IL: Ionic liquid; PS: Polystyrene; SBA-15: Mesoporous silica type SBA-15; [2] 2D GC (heart-cut MDGC); [3] Comprehensive 2D GC (GCxGC).

### 3. Sample Matrices

Although the determination of pesticides in multiple matrices may be required, the use of SPME as sample pretreatment seems to be focused on environmental and food samples to date. Nevertheless, the determination of pesticides using SPME has also been applied to other matrices, such as human fluids [73–75] or agricultural non-food products, such as textiles [76]. Although these applications have shown very promising results, the application of SPME in these areas remains limited when compared to environmental and food applications. The particular interest in these fields is likely due to the regulations imposed by different authorities across the world. In general, pesticides have maximum residue levels allowable in the order of ppb and, even if many of these compounds may be included, no listing can be exhaustive as new variants may be developed and applied despite the local regulations. Therefore, this is a continuously evolving field. SPME has the great advantage of extracting the analytes from the matrix without diluting them, making it possible to reach the needed detection limits, as can be seen in Table 1. Nevertheless, of note, this entails the risk of falsely disregarding some analytes that are not correctly extracted from the matrix by this means. In the following sections, we discuss the applications in food and environmental samples, focusing on the main analytical strategies to attain the required sensitivity.

*3.1. Food Samples*

Agricultural practices had always evolved to cope with the constantly increasing demand for food. The use of pesticides to improve plant growth and pest control have been a crucial development. However, the widespread use of these products entails some potential risks for the environment and health, in long and short terms. Indeed, when pesticides are misused, food products can be a carrier of pesticides into the human body [77]. For this reason, stringent regulations have been in place for some time now [78]. Given that the allowed limits for these contaminants usually are at the ppb level and that food samples are extremely complex matrices, the sample pre-treatment has been a key parameter in the development of new analytical methods in this field. Although classic extraction methods, such as LLE are still in use, they are often combined in multiple steps to enrich the extracts in the analytes of interest effectively. Therefore, techniques that can be very selective in the extraction process have become particularly interesting in this field to reduce matrix interference. This is the case for SPME, in which its potential in this area has been largely explored since its introduction in the early 1990s [79].

In this area, SPME has been implemented in both modes, HS-SPME and DI-SPME. However, the direct immersion mode benefits from higher recoveries, becoming the preferred choice in food matrices. Liquid samples, such as juices [39,53], wines [53] or milk [40,54], can be easily analyzed by DI-SPME. However, they may require some further pre-treatment prior to extraction to minimize matrix effects. For instance, Gonzalez-Rodriguez et al. [40] found poor repeatability and sensitivity when applying DI-SPME to untreated and non-diluted milk samples, which is likely related to the high lipid and protein content of the sample. Fortunately, a simple dilution in water sufficed to overcome the issue. In the case of juices, the reported sampling conditions largely vary between applications. For example, Zambonin et al. [53] reported simple centrifugation and dilution with water prior to DI-SPME extraction, while Cortes-Aguado et al. [39] performed an extraction in ethyl acetate, followed by centrifugation, evaporation to dryness, and dissolution in water:acetone mixture prior to DI-SPME extraction. Therefore, sample pre-treatments and extraction conditions should always be optimized for the intended application.

Other samples, such as vegetables and fruits [36,42,44–48] or honey [38,56] require blending and mixing with diluents to allow for the migration of analytes to the sorbent. Vegetables and fruits are blended and diluted, and in some cases, a centrifugation step is added to further clean-up the matrix. Moreover, in these cases, the extraction conditions vary between applications despite the sample nature. In some cases, the use of small quantities of solvent mixtures are reported [36,42,44,45], while in others, a centrifugation step is included to further clean-up the sample [46–48]. Of note, the optimal dilution medium may be different for different analytes even within the same matrix, as reported by Menezes Filho et al. [36]. In the case of honey samples, pre-treatments largely vary without justification, analytes were simply extracted by immersing the SPME fiber into heated buffered aqueous sample solutions of the sample [56] or using a multistep clean-up procedure, including solvent extraction (with small solvent amounts), centrifugation, and concentration by evaporation and redissolution in solvent [38]. Although in the second case 60 analytes were studied (vs. 16 in the other case), the clean-up steps prior to DI-SPME are more cumbersome. Therefore, the pre-treatment should be carefully considered depending on the intended application.

Despite the sample under study, clean-up procedures prior to DI-SPME extraction and extraction conditions require careful consideration. Therefore, parameters, such as pH, temperature, time, and ionic strength are commonly explored during optimization. Of note, the effect of the ionic strength does not always have a significant benefit for the extraction process when working in DI-SPME mode [39,40]. Indeed, the sensitivity may be increased (up to a specific ionic strength) for some analytes by the salting-out effect, while it may be reduced for others. Although samples may require a clean-up step prior to the extraction by DI-SPME, it should be noted that this step remains less labor intensive than conventional techniques, such as LLE. Moreover, even when organic solvents are needed, they are used in very small amounts compared to other techniques, making DI-SPME a greener alternative.

In the case of HS-SPME, the matrix contamination is further minimized and the obtained chromatograms tend to be less complex. Although this mode limits the pesticides that can be analyzed to those that present good affinity toward the gas phase, the sensitivity of these analytes is increased, as observed by Menezes Filho et al. [36]. The same variety of samples as reported for DI-SPME are reported using HS-SPME, namely, fruits and vegetables [35,36,49–51], juices [41], honey [57], and milk [37,40,54]. The sample pre-treatment is not significantly different from the ones for DI-SPME. Fruits and vegetables are first homogenized by blending, then centrifugation and filtration may be applied to further clean-up the sample, although this is less needed than for DI-SPME. Thereafter, in all samples, usually, a small amount of solvents, including a brine solution, is added prior to HS-SPME extraction. Of note, this may not always be the best approach, for instance, Rodrigues et al. [37] found that the salting-out effect was detrimental in the recovery of organophosphorus pesticides in milk.

The HS-SPME extraction is usually carried out by heating the sample to release the analytes from the matrix to the HS. Therefore, the extraction temperature and time have become fundamental parameters in this case, and require careful optimization. It is not surprising to find rising edge situations in response surfaces toward higher temperatures [37]. However, there is always a physical limitation of the maximum temperature a sample may withhold prior to degradation. Although using vacuum as an alternative has been proposed in the area [80], its use is not widespread, and to the best of our knowledge, it has not been reported for pesticide analysis in food matrices.

Due to the highly diverse nature of food products, the extraction procedure optimization must be carried out in a case-by-case approach. Therefore, the comparison of different SPME fiber coatings that are commercially available [35,36,38,40,42,48,54,56] or that are specifically developed for a certain application [44,47,50,57] is often reported. Moreover, the use of experimental designs to obtain optimal conditions has gained popularity, which is likely thanks to the simple to use of software to process the data. Using response surface models after exploring the effects of different extraction parameters, such as temperature, time, pH, ionic strength, etc. has become a common practice [35,37,44,46,48,49,54].

Whether it is DI-SPME or HS-SPME, GC is largely chosen as the analytical separation technique. This is mostly thanks to its simplicity for automation when using SPME fibers and the good compatibility with the analytes that are extracted by SPME. Indeed, thermal desorption is largely used to bring the sample into the system thanks to the simplicity and straight compatibility of the SPME fibers with the splitless injector. On the other end of the GC, depending on the target analytes and the application possibilities, different detection techniques are employed. Due to the low levels that need to be detected, ECD [47,50,51,54] and MS [35–42,44–46,48,52,53,57] are the most reported detection techniques for this type of samples. Regarding the calibration strategy for quantification, it has to be emphasized that it needs to account for the complex and diverse matrices found in food samples. Standard mixtures for the optimization and validation of the methods, and for quantitation are easily available in set mixtures. These standard mixtures can be used to spike samples of interest or blank samples. In many cases, food samples from organic farming are considered to be free of pesticides and are used as matrix blanks [36–38,40,44,45,47,48,50,51]. However, these are not certified materials (at least not for research purposes) and they should be screened to confirm that they are really pesticide free. As aforementioned, the use of multidimensional analysis, such as GC×GC, could be helpful for these untargeted screenings.

*3.2. Environmental Samples*

The rapid increase in the use of pesticides for agricultural practices produced a markedly social concern about the levels of these active ingredients in the environment. As a consequence, the residual levels allowed for these substances are being regulated toward values extremely lower in several matrices, including environmental ones, which include water and soil. Along with these regulations, there is a high demand for analytical methodologies for accurate pesticide determination at trace concentration levels, which overcome the disadvantages of traditional procedures. Namely, long preparation time and large quantities of solvent consumption, as well as laboratory-generated waste. In this regard, SPME has gained special attention and its combination with chromatographic techniques hyphenated to mass spectrometry has been widely used for identification and quantification due to both mentioned pretreatment and separative techniques providing significant improvement in sensitivity and selectivity. Therefore, more extensive research in the area is highly encouraged.

Soil and sediment samples represent challenging matrices since their composition includes a wide variety of minerals, humidity content, and organic matter. The latter acts as a sorbent to which analytes are strongly bonded, hindering the extractive process. As a result, classical method approaches, including Soxhlet extraction, supercritical fluid extraction, accelerated solvent extraction, ultrasonic extraction, and microwave-assisted extraction were selected primarily as pretreatment procedures to desorb analytes from these types of samples. Fortunately, since their introduction, SPME has grown notably and advances related to sorbent development, automation, and SPME-assisted technologies allowed the technique to be considered as an alternative and improved tool for soil and sediment analysis. In these matrices, both DI and HS approaches have been reported for the determination of pesticides and mainly for other environmental contaminants [52], considering two protocols: (i) DI of the fiber in the sample and (ii) preparation of a suspension of the solid matrix by addition of water, brine or a mixture of solvents followed by SPME fiber exposure to the HS or by DI into the liquid phase. On the one hand, the first approach avoids the use of solvent, but prevents the correct diffusion of analytes to the sorbent due to the impossibility of agitating. As a result, microextraction may not be representative of the whole sample. On the other hand, the second strategy provides better extraction efficiencies since the suspension aids in the analyte release from sample pores and allows homogenization by agitation. In this approach, HS is preferred compared to DI since the lifespan of fiber is prolonged.

Even though the application of SPME in soil sampling focused on pesticide determination is scarce, there are some examples in the literature [52,58,59,72,81,82] covering general workflows on which most reported methods are based. For example, Bouaid et al. [58] reported a method based on SPME for the determination of atrazine and four organophosphorus pesticides (parathion-methyl, chlorpyriphos, methidation, and carbophention) in sandy soil samples. Different experimental variables (extraction time, concentration of sodium chloride solution, and desorption time and temperature) were optimized using a central composite design. The final procedure consisted of a previous extraction of the analytes from a weighed amount of soil sample with a small volume of organic solvent (methanol) subjected to ultrasonic agitation. After that extraction, solid particles were separated from the supernatant by centrifugation. A small aliquot of the extract was diluted with deionized water, and this aqueous solution was used for SPME after the addition of

NaCl to increase extraction efficiency by the salting out effect. SPME microextraction was performed at ambient temperature by DI of a 100 µm PDMS fiber into the solution under magnetic stirring. Extracted compounds were introduced into a GC by thermal desorption in a splitless operated injector. Analyses were performed in two parallel systems, using an ECD in one case and MS detection in the other. The MS was used in SIM mode, with two fragments for each compound for confirmation. It should be noted that operating the MS in this way solved interferences observed when using the ECD. However, as only two ions were monitored, no confident identification would have been possible [83]. Limits of detection achieved in real soil samples were in the range of 0.6–7 ng g$^{-1}$ except for atrazine and methyl parathion which were 30 and 2 ng g$^{-1}$, respectively.

Similarly, Chang et al. [72] developed a HS-SPME method to determine 10 OCPs in surface estuarine sediments. The optimized procedure consisted of preparing a slurry by placing a weighed amount of sediments previously sieved, deionized water containing a surfactant, and a magnetic bar into a sealed vial. Thereafter, SPME was performed at 70 °C by exposure of 100 µm PDMS fiber to the HS for 60 min while continuously agitating the slurry. Finally, desorption was conducted thermally in a GC injector. Separation was carried out in a gas chromatograph equipped with an ECD. Quantification was performed by analysis of aqueous standards over the range of 0.2–4 ng g$^{-1}$. The developed procedure was compared with Soxhlet extraction using a certified sample, demonstrating good analytical performance and a clear agreement between both methods.

Regarding water samples, the composition may be as diverse as soil samples, depending on its origin and location. Variable amounts of dissolved organic matter, presence of solid particles, suspended sediments, non-aqueous phase liquids, as well as dissolved gases and inorganic ions may be present. This variability can be attributed to the diversity of water samples which includes groundwater; river, lake, and seawater; influent and effluent wastewater; tap water; ice cores; and snow samples. Even though drinking water may be considered as a food sample, it is reasonable to include it in this section due to matrix similarity. Official methods for isolation and/or preconcentration of target pesticides and other pollutants are based on LLE and SPE [84,85]. In addition to the abovementioned drawbacks related to the multiples stages, time, and solvent consumption, LLE has the disadvantages of being laborious, requiring additional steps for clean-up and concentration by evaporation, and often lead to emulsion occurrence due to the presence of surfactants in the sample. Regarding the SPE approach, negative aspects rely on the need for previous stages of centrifugation or filtration to avoid cartridge clogging. In addition, SPE capacity may not be sufficient, yielding insufficient sensitivity. Therefore, it is not surprising that SPME appears as a powerful and improved technique for sample pretreatment. In this area, a vast number of publications are available [58–66,68–71] dealing with many classes of pesticides in different types of samples. Moreover, non-commercially available fibers with new coating materials were developed for these applications. As presented in Table 1, the results showed good performances with very low LOD in the range of ng L$^{-1}$ or below. Quantification in water analysis by SPME is usually carried out by external calibration prepared with deionized water spiked with pesticides standard solutions and extracting them with the same procedure as the sample. In the majority of the reviewed papers, the determination of pesticides is carried out by GC introducing extracted analytes by thermal desorption, employing mainly MS, ECD, and NPD detection, although other detection systems, such as FID have also been used.

Concerning the SPME mode, HS or DI modes have been reported. For example, Gong et al. [65] have recently reported the fabrication of a novel SPME fiber based on NU-1000 (a zirconium-based metal-organic framework) to be employed for DI extraction of seven OCPs applied to pond and river water. The analytical performance of this fiber coupled with GC-MS (SIM mode) was evaluated under optimal conditions. Briefly, the methodology consisted of filtering water samples and immersing the fiber into a sample aliquot maintained under vibration. DI-SPME optimized conditions were extraction time of 30 min, extraction temperature of 40 °C, desorption time and temperature of 6 min and 260 °C, respectively. Furthermore, the developed fiber was compared with 65 um PDMS/DVB and 85 μm PA fibers commercially available fibers yielding extraction efficiencies that were 3–10 and 2–20 times higher, respectively.

Domínguez et al. [71] developed a method to simultaneously determine 15 pesticides and other commonly found pollutants in water samples at ultra-trace levels, such as polycyclic aromatic hydrocarbons, polychlorinated biphenyls, and brominated diphenyl ethers. The proposed method based on a combination of HS-SPME and gas chromatography was coupled to high-resolution mass spectrometry in wastewater samples. The authors optimized the extraction procedure and calculated LOQs from matrix-matched calibrations, ranging from 0.01 to 350 ng $L^{-1}$. This approach allowed an increase in the selectivity of the extraction method, largely reducing the potential interferences caused by high molecular mass and other non-volatile interferences present in the sample matrix. At the same time, the lack of direct contact between the fiber and the sample helped in protecting the fiber coating from damage and extended the lifetime of fibers.

Even though HS-SPME is preferable over DI-SPME to prevent fiber damage, it is worth mentioning that an additional step to release analytes from the sample to the HS has to be incorporated prior to uptake by the fiber coating. Consequently, an extra equilibrium is involved in the overall process and extraction efficiencies may be impaired, requiring, therefore, strategies to achieve analyte transport to the HS in operatively reasonable times, especially for analytes with low Henry's constants. To overcome this limitation, strategies including cold-fiber SPME approach [21,86], vacuum-assisted HS-SPME [86,87], as well as ultrasonic [88], microwave-assisted HS-SPME [89] or multiple-cumulative trapping SPME [90,91] could be applied.

## 4. Concluding Remarks

The analysis of water and food samples to control the presence of pesticides for environmental monitoring or to ensure that products are safe for consumption has become of major interest. Considering the large number of compounds that need to be tested, multiresidue and untargeted pesticides analysis are fundamental to ensure an appropriate determination of all pesticides that can be present. Otherwise, some analytes may be wrongly disregarded, creating a potential environmental and health risk. Given the high complexity and variety of samples, pre-treatment techniques are needed to deal with the complex matrices. SPME has gained the attention of researchers for its simplicity and efficiency in these cases. Indeed, SPME requires only a minimal sample quantity with a simple dilution with small amounts of solvents, often aqueous, to extract multiple analytes of interest in a single-step extraction, making it also a green analytical technique. Moreover, using fibers makes it easy to handle, and the whole process can be automated, contributing to its acceptance for routine applications. Nevertheless, as the sample matrix is almost unique in each case, optimization is required for each sample and different coatings should be tested to find the best solid phase material. On the one hand, this gives rise to new materials that can improve the extraction capacity and be more specifically engineered for these applications. On the other hand, the design of experiments for the optimization of the extraction procedure should be implemented to obtain meaningful and reproducible results. Moreover, different solid phases should be tested, as some analytes may not be adequately extracted due to poor recoveries with certain phases.

As pesticides are a large variety of molecules, multiresidue screening is required, even untargeted analysis should be considered to avoid falsely disregarding some pesticides. Therefore, the use of MS for identification has become unavoidable. Nevertheless, even after SPME extraction, these analyses are targeted and MS is used as an extra separation dimension due to the complexity of the chromatograms obtained and the low detection limits required, losing the full MS identification capacity as it is necessary to rely on confirmation that is derived from the ratio of specific fragments. Therefore, the high separation power of multidimensional chromatography, along with the increased sensitivity obtained thanks to the band compression that occurs in the modulator, could be an interesting approach to obtain better resolved chromatograms that allow for the use of MS in scan mode to identify the different components. Unfortunately, only a very limited number of publications explored this possibility, although its potential has been made evident in other fields.

In conclusion, we hypothesize that the need for pesticides analysis will continue to grow and that SPME and GC will continue to play a central role. New and more efficient materials for SPME may be developed and the use of the design of experiments for optimization of the extraction procedure should be encouraged along with the development of new SPME extraction strategies. Finally, the great potential benefit of multidimensional chromatography, and, in particular, comprehensive techniques, such as GC×GC, should be considered to overcome the challenges posed by the need for a reliable multiresidue screening analysis.

**Author Contributions:** Conceptualization, J.A., C.L. and G.P.; writing—original draft preparation, J.A. and C.L.; writing—review and editing, J.A., C.L. and G.P.; supervision, G.P.; project administration, G.P.; funding acquisition, G.P. All authors have read and agreed to the published version of the manuscript.

**Funding:** This work received no external funding.

**Institutional Review Board Statement:** Not applicable.

**Informed Consent Statement:** Not applicable.

**Data Availability Statement:** Not applicable.

**Acknowledgments:** This article is based on work from the Sample Preparation Study Group and Network, supported by the Division of Analytical Chemistry of the European Chemical Society.

**Conflicts of Interest:** The authors declare no conflict of interest.

## References

1. Arthur, C.L.; Pawliszyn, J. Solid Phase Microextraction with Thermal Desorption Using Fused Silica Optical Fibers. *Anal. Chem.* **1990**, *62*, 2145–2148. [CrossRef]
2. Panighel, A.; Flamini, R. Applications of Solid-Phase Microextraction and Gas Chromatography/Mass Spectrometry (SPME-GC/MS) in the Study of Grape and Wine Volatile Compounds. *Molecules* **2014**, *19*, 21291–21309. [CrossRef] [PubMed]
3. Riboni, N.; Fornari, F.; Bianchi, F.; Careri, M. Recent Advances in in Vivo Spme Sampling. *Separations* **2020**, *7*, 6. [CrossRef]
4. Moreira, N.; Lopes, P.; Cabral, M.; Guedes de Pinho, P. HS-SPME/GC-MS Methodologies for the Analysis of Volatile Compounds in Cork Material. *Eur. Food Res. Technol.* **2016**, *242*, 457–466. [CrossRef]
5. Kataoka, H.; Saito, K. Recent Advances in SPME Techniques in Biomedical Analysis. *J. Pharm. Biomed. Anal.* **2011**, *54*, 926–950. [CrossRef]
6. Pati, S.; Tufariello, M.; Crupi, P.; Coletta, A.; Grieco, F.; Losito, I. Quantification of Volatile Compounds in Wines by HS-SPME-GC/MS: Critical Issues and Use of Multivariate Statistics in Method Optimization. *Processes* **2021**, *9*, 662. [CrossRef]
7. Ouyang, G.; Pawliszyn, J. SPME in Environmental Analysis. *Anal. Bioanal. Chem.* **2006**, *386*, 1059–1073. [CrossRef]
8. Souza-Silva, É.A.; Jiang, R.; Rodríguez-Lafuente, A.; Gionfriddo, E.; Pawliszyn, J. A Critical Review of the State of the Art of Solid-Phase Microextraction of Complex Matrices I. Environmental Analysis. *TrAC Trends Anal. Chem.* **2015**, *71*, 224–235. [CrossRef]
9. de Fátima Alpendurada, M. Solid-Phase Microextraction: A Promising Technique for Sample Preparation in Environmental Analysis. *J. Chromatogr. A* **2000**, *889*, 3–14. [CrossRef]
10. Environmental Protection Agency What Is a Pesticide? Available online: https://www.epa.gov/minimum-risk-pesticides/what-pesticide (accessed on 28 August 2022).
11. Hassaan, M.A.; El Nemr, A. Pesticides Pollution: Classifications, Human Health Impact, Extraction and Treatment Techniques. *Egypt. J. Aquat. Res.* **2020**, *46*, 207–220. [CrossRef]
12. Narenderan, S.T.; Meyyanathan, S.N.; Babu, B. Review of Pesticide Residue Analysis in Fruits and Vegetables. Pre-Treatment, Extraction and Detection Techniques. *Food Res. Int.* **2020**, *133*, 109141. [CrossRef]
13. Rani, L.; Thapa, K.; Kanojia, N.; Sharma, N.; Singh, S.; Grewal, A.S.; Srivastav, A.L.; Kaushal, J. An Extensive Review on the Consequences of Chemical Pesticides on Human Health and Environment. *J. Clean. Prod.* **2021**, *283*, 124657. [CrossRef]
14. Fenik, J.; Tankiewicz, M.; Biziuk, M. Properties and Determination of Pesticides in Fruits and Vegetables. *TrAC-Trends Anal. Chem.* **2011**, *30*, 814–826. [CrossRef]
15. Li, Z.; Jennings, A. Worldwide Regulations of Standard Values of Pesticides for Human Health Risk Control: A Review. *Int. J. Environ. Res. Public Health* **2017**, *14*, 826. [CrossRef]

16. Tobiszewski, M.; Mechlinska, A.; Namie, J. Green Analytical Chemistry—Theory and Practice. *Chem. Soc. Rev.* **2010**, *39*, 2869–2878. [CrossRef]
17. Armenta, S.; Garrigues, S.; Esteve-Turrillas, F.A.; de la Guardia, M. Green Extraction Techniques in Green Analytical Chemistry. *TrAC - Trends Anal. Chem.* **2019**, *116*, 248–253. [CrossRef]
18. Elsevier Scopus. Available online: https://www.scopus.com/results/results.uri?sort=plf-f&src=s&st1=pesticides&st2=gas+chromatography&searchTerms=SPME%3f%21%22*%24&sid=b6292b113259b87b3290315e5d5d5364&sot=b&sdt=b&sl=89&s=%28TITLE-ABS-KEY%28pesticides%29+AND+TITLE-ABS-KEY%28gas+chromatography%29+AND+TITLE-ABS-KEY%28SPME%29%29&origin=searchbasic&editSaveSearch=&yearFrom=Before+1960&yearTo=Present (accessed on 28 August 2022).
19. Zhang, Q.H.; Zhou, L.D.; Chen, H.; Wang, C.Z.; Xia, Z.N.; Yuan, C.S. Solid-Phase Microextraction Technology for in Vitro and in Vivo Metabolite Analysis. *TrAC-Trends Anal. Chem.* **2016**, *80*, 57–65. [CrossRef]
20. Reyes-Garcés, N.; Gionfriddo, E.; Gómez-Ríos, G.A.; Alam, M.N.; Boyacl, E.; Bojko, B.; Singh, V.; Grandy, J.; Pawliszyn, J. Advances in Solid Phase Microextraction and Perspective on Future Directions. *Anal. Chem.* **2018**, *90*, 302–360. [CrossRef]
21. Pawliszyn, J. Theory of Solid-Phase Microextraction. In *Handbook of Solid Phase Microextraction*; Elsevier: Amsterdam, The Netherlands, 2012; pp. 13–59. ISBN 9780124160170.
22. Lancioni, C.; Castells, C.; Candal, R.; Tascon, M. Headspace Solid-Phase Microextraction: Fundamentals and Recent Advances. *Adv. Sample Prep.* **2022**, *3*, 100035. [CrossRef]
23. Gómez-Ríos, G.A.; Garcés, N.R.; Tascon, M. Smart Materials in Solid Phase Microextraction (SPME). In *Handbook of Smart Materials in Analytical Chemistry*; John Wiley & Sons, Ltd.: Chichester, UK, 2019; pp. 581–620.
24. Patel, D.I.; Roychowdhury, T.; Shah, D.; Jacobsen, C.; Herrington, J.S.; Hoisington, J.; Myers, C.; Salazar, B.G.; Walker, A.V.; Bell, D.S.; et al. 6-Phenylhexyl Silane Derivatized, Sputtered Silicon Solid Phase Microextraction Fiber for the Parts-per-trillion Detection of Polyaromatic Hydrocarbons in Water and Baby Formula. *J. Sep. Sci.* **2021**, *44*, 2824–2836. [CrossRef]
25. Dugheri, S.; Mucci, N.; Cappelli, G.; Trevisani, L.; Bonari, A.; Bucaletti, E.; Squillaci, D.; Arcangeli, G. Advanced Solid-Phase Microextraction Techniques and Related Automation: A Review of Commercially Available Technologies. *J. Anal. Methods Chem.* **2022**, *2022*, 72. [CrossRef]
26. David, F.; Ochiai, N.; Sandra, P. Two Decades of Stir Bar Sorptive Extraction: A Retrospective and Future Outlook. *TrAC Trends Anal. Chem.* **2019**, *112*, 102–111. [CrossRef]
27. Herrington, J.S.; Gómez-Ríos, G.A.; Myers, C.; Stidsen, G.; Bell, D.S. Hunting Molecules in Complex Matrices with Spme Arrows: A Review. *Separations* **2020**, *7*, 12. [CrossRef]
28. Jiang, R.; Pawliszyn, J. Thin-Film Microextraction Offers Another Geometry for Solid-Phase Microextraction. *TrAC Trends Anal. Chem.* **2012**, *39*, 245–253. [CrossRef]
29. Piri-Moghadam, H.; Gionfriddo, E.; Grandy, J.J.; Alam, M.N.; Pawliszyn, J. Development and Validation of Eco-Friendly Strategies Based on Thin Film Microextraction for Water Analysis. *J. Chromatogr. A* **2018**, *1579*, 20–30. [CrossRef]
30. Merkle, S.; Kleeberg, K.; Fritsche, J. Recent Developments and Applications of Solid Phase Microextraction (SPME) in Food and Environmental Analysis—A Review. *Chromatography* **2015**, *2*, 293–381. [CrossRef]
31. Raeppel, C.; Fabritius, M.; Nief, M.; Appenzeller, B.M.R.; Millet, M. Coupling ASE, Sylilation and SPME–GC/MS for the Analysis of Current-Used Pesticides in Atmosphere. *Talanta* **2014**, *121*, 24–29. [CrossRef]
32. Henriksen, T.; Svensmark, B.; Lindhardt, B.; Juhler, R.K. Analysis of Acidic Pesticides Using in Situ Derivatization with Alkylchloroformate and Solid-Phase Microextraction (SPME) for GC–MS. *Chemosphere* **2001**, *44*, 1531–1539. [CrossRef]
33. Pena-Pereira, F. *Miniaturization in Sample Preparation*; De Gruyter Open: Berlin, Germany, 2014; ISBN 978-3-11-041017-4.
34. López-Lorente, Á.I.; Pena-Pereira, F.; Pedersen-Bjergaard, S.; Zuin, V.G.; Ozkan, S.A.; Psillakis, E. The Ten Principles of Green Sample Preparation. *TrAC-Trends Anal. Chem.* **2022**, *148*. [CrossRef]
35. Abdulra'uf, L.B.; Tan, G.H. Chemometric Approach to the Optimization of HS-SPME/GC–MS for the Determination of Multiclass Pesticide Residues in Fruits and Vegetables. *Food Chem.* **2015**, *177*, 267–273. [CrossRef]
36. Menezes Filho, A.; dos Santos, F.N.; de Paula Pereira, P.A. Development, Validation and Application of a Methodology Based on Solid-Phase Micro Extraction Followed by Gas Chromatography Coupled to Mass Spectrometry (SPME/GC–MS) for the Determination of Pesticide Residues in Mangoes. *Talanta* **2010**, *81*, 346–354. [CrossRef]
37. Rodrigues, F.d.M.; Mesquita, P.R.R.; de Oliveira, L.S.; de Oliveira, F.S.; Menezes Filho, A.; de P. Pereira, P.A.; de Andrade, J.B. Development of a Headspace Solid-Phase Microextraction/Gas Chromatography–Mass Spectrometry Method for Determination of Organophosphorus Pesticide Residues in Cow Milk. *Microchem. J.* **2011**, *98*, 56–61. [CrossRef]
38. Al-Alam, J.; Fajloun, Z.; Chbani, A.; Millet, M. A Multiresidue Method for the Analysis of 90 Pesticides, 16 PAHs, and 22 PCBs in Honey Using QuEChERS–SPME. *Anal. Bioanal. Chem.* **2017**, *409*, 5157–5169. [CrossRef]
39. Cortés-Aguado, S.; Sánchez-Morito, N.; Arrebola, F.J.; Frenich, A.G.; Vidal, J.L.M. Fast Screening of Pesticide Residues in Fruit Juice by Solid-Phase Microextraction and Gas Chromatography-Mass Spectrometry. *Food Chem.* **2008**, *107*, 1314–1325. [CrossRef]
40. González-Rodríguez, M.J.; Arrebola Liébanas, F.J.; Garrido Frenich, A.; Martínez Vidal, J.L.; Sánchez López, F.J. Determination of Pesticides and Some Metabolites in Different Kinds of Milk by Solid-Phase Microextraction and Low-Pressure Gas Chromatography-Tandem Mass Spectrometry. *Anal. Bioanal. Chem.* **2005**, *382*, 164–172. [CrossRef]
41. Del Castillo, M.L.R.; Rodriguez-Valenciano, M.; De La Peña Moreno, F.; Blanch, G.P. Evaluation of Pesticide Residue Contents in Fruit Juice by Solid-Phase Microextraction and Multidimensional Gas Chromatography Coupled with Mass Spectrometry. *Talanta* **2012**, *89*, 77–83. [CrossRef]

42. Ruiz del Castillo, M.L.; Rodríguez-Valenciano, M.; Flores, G.; Blanch, G.P. New Method Based on Solid Phase Microextraction and Multidimensional Gas Chromatography-Mass Spectrometry to Determine Pesticides in Strawberry Jam. *LWT-Food Sci. Technol.* **2019**, *99*, 283–290. [CrossRef]
43. Purcaro, G.; Quinto Tranchida, P.; Conte, L.; Obiedzińska, A.; Dugo, P.; Dugo, G.; Mondello, L. Performance Evaluation of a Rapid-Scanning Quadrupole Mass Spectrometer in the Comprehensive Two-Dimensional Gas Chromatography Analysis of Pesticides in Water. *J. Sep. Sci.* **2011**, *34*, 2411–2417. [CrossRef]
44. Pang, Y.; Zang, X.; Li, H.; Liu, J.; Chang, Q.; Zhang, S.; Wang, C.; Wang, Z. Solid-Phase Microextraction of Organophosphorous Pesticides from Food Samples with a Nitrogen-Doped Porous Carbon Derived from g-C3N4 Templated MOF as the Fiber Coating. *J. Hazard. Mater.* **2020**, *384*, 121430. [CrossRef]
45. Sapahin, H.A.; Makahleh, A.; Saad, B. Determination of Organophosphorus Pesticide Residues in Vegetables Using Solid Phase Micro-Extraction Coupled with Gas Chromatography–Flame Photometric Detector. *Arab. J. Chem.* **2015**, *12*, 1934–1944. [CrossRef]
46. Souza-Silva, É.A.; Pawliszyn, J. Direct Immersion Solid-Phase Microextraction with Matrix-Compatible Fiber Coating for Multiresidue Pesticide Analysis of Grapes by Gas Chromatography–Time-of-Flight Mass Spectrometry (DI-SPME-GC-ToFMS). *J. Agric. Food Chem.* **2015**, *63*, 4464–4477. [CrossRef] [PubMed]
47. Wang, M.; Zhou, X.; Zang, X.; Pang, Y.; Chang, Q.; Wang, C.; Wang, Z. Determination of Pesticides Residues in Vegetable and Fruit Samples by Solid-Phase Microextraction with a Covalent Organic Framework as the Fiber Coating Coupled with Gas Chromatography and Electron Capture Detection. *J. Sep. Sci.* **2018**, *41*, 4038–4046. [CrossRef] [PubMed]
48. Zhang, L.; Gionfriddo, E.; Acquaro, V.; Pawliszyn, J. Direct Immersion Solid-Phase Microextraction Analysis of Multi-Class Contaminants in Edible Seaweeds by Gas Chromatography-Mass Spectrometry. *Anal. Chim. Acta* **2018**, *1031*, 83–97. [CrossRef] [PubMed]
49. Delinska, K.; Yavir, K.; Kloskowski, A. Head-Space SPME for the Analysis of Organophosphorus Insecticides by Novel Silica IL-Based Fibers in Real Samples. *Molecules* **2022**, *27*, 4688. [CrossRef]
50. Wu, M.; Chen, G.; Ma, J.; Liu, P.; Jia, Q. Fabrication of Cross-Linked Hydrazone Covalent Organic Frameworks by Click Chemistry and Application to Solid Phase Microextraction. *Talanta* **2016**, *161*, 350–358. [CrossRef]
51. Mee Kin, C.; Guan Huat, T. Headspace Solid-Phase Microextraction for the Evaluation of Pesticide Residue Contents in Cucumber and Strawberry after Washing Treatment. *Food Chem.* **2010**, *123*, 760–764. [CrossRef]
52. Boyd-boland, A.A.; Pawliszyn, J.B. Solid-Phase Microextraction of Nitrogen-Containing Herbicides. *J. Chromatogr. A* **1995**, *704*, 163–172. [CrossRef]
53. Zambonin, C.G.; Quinto, M.; De Vietro, N.; Palmisano, F. Solid-Phase Microextraction–Gas Chromatography Mass Spectrometry: A Fast and Simple Screening Method for the Assessment of Organophosphorus Pesticides Residues in Wine and Fruit Juices. *Food Chem.* **2004**, *86*, 269–274. [CrossRef]
54. Fernandez-Alvarez, M.; Llompart, M.; Lamas, J.P.; Lores, M.; Garcia-Jares, C.; Cela, R.; Dagnac, T. Development of a Solid-Phase Microextraction Gas Chromatography with Microelectron-Capture Detection Method for a Multiresidue Analysis of Pesticides in Bovine Milk. *Anal. Chim. Acta* **2008**, *617*, 37–50. [CrossRef]
55. Akbarzade, S.; Chamsaz, M.; Rounaghi, G.H.; Ghorbani, M. Zero Valent Fe-Reduced Graphene Oxide Quantum Dots as a Novel Magnetic Dispersive Solid Phase Microextraction Sorbent for Extraction of Organophosphorus Pesticides in Real Water and Fruit Juice Samples Prior to Analysis by Gas Chromatography-Mass Spectrom. *Anal. Bioanal. Chem.* **2018**, *410*, 429–439. [CrossRef]
56. Campillo, N.; Peñalver, R.; Aguinaga, N.; Hernández-Córdoba, M. Solid-Phase Microextraction and Gas Chromatography with Atomic Emission Detection for Multiresidue Determination of Pesticides in Honey. *Anal. Chim. Acta* **2006**, *562*, 9–15. [CrossRef]
57. Zali, S.; Jalali, F.; Es-haghi, A.; Shamsipur, M. Electrospun Nanostructured Polystyrene as a New Coating Material for Solid-Phase Microextraction: Application to Separation of Multipesticides from Honey Samples. *J. Chromatogr. B* **2015**, *1002*, 387–393. [CrossRef] [PubMed]
58. Bouaid, A.; Ramos, L.; Gonzalez, M.; Fernández, P.; Cámara, C. Solid-Phase Microextraction Method for the Determination of Atrazine and Four Organophosphorus Pesticides in Soil Samples by Gas Chromatography. *J. Chromatogr. A* **2001**, *939*, 13–21. [CrossRef]
59. Boyd-boland, A.A.; Magdic, S.; Pawliszyn, J.B. Simultaneous Determination of 60 Pesticides in Water Using Solid-Phase Microextraction and Gas Chromatography-Mass Spectrometry. *Analyst* **1996**, *121*, 929–938. [CrossRef]
60. Beceiro-González, E.; Concha-Graña, E.; Guimaraes, A.; Gonçalves, C.; Muniategui-Lorenzo, S.; Alpendurada, M.F. Optimisation and Validation of a Solid-Phase Microextraction Method for Simultaneous Determination of Different Types of Pesticides in Water by Gas Chromatography-Mass Spectrometry. *J. Chromatogr. A* **2007**, *1141*, 165–173. [CrossRef]
61. Carabias-Martínez, R.; García-Hermida, C.; Rodríguez-Gonzalo, E.; Ruano-Miguel, L. Behaviour of Carbamate Pesticides in Gas Chromatography and Their Determination with Solid-Phase Extraction and Solid-Phase Microextraction as Preconcentration Steps. *J. Sep. Sci.* **2005**, *28*, 2130–2138. [CrossRef]
62. Eisert, R.; Levsen, K. Determination of Pesticides in Aqueous Samples by Solid-Phase Microextraction in-Line Coupled to Gas Chromatography—Mass Spectrometry. *J. Am. Soc. Mass Spectrom.* **1995**, *6*, 1119–1130. [CrossRef]
63. Elizarragaz-de la Rosa, D.; Guzmán-Mar, J.L.; Salas-Espinosa, E.A.; Heras-Ramírez, M.E.; Hinojosa-Reyes, L.; Gaspar-Ramírez, O.; Ruiz-Ruiz, E.J. Multi-Residual Determination of Multi-Class Pesticides in Groundwater by Direct Immersion Solid-Phase Microextraction with Gas Chromatography-Selected Ion Monitoring Mass Spectrometry (GC–MS/SIM) Detection. *Water Air Soil Pollut.* **2022**, *233*, 76. [CrossRef]

64. Gonçalves, C.; Alpendurada, M.F. Solid-Phase Micro-Extraction-Gas Chromatography-(Tandem) Mass Spectrometry as a Tool for Pesticide Residue Analysis in Water Samples at High Sensitivity and Selectivity with Confirmation Capabilities. *J. Chromatogr. A* **2004**, *1026*, 239–250. [CrossRef]

65. Gong, X.; Xu, L.; Huang, S.; Kou, X.; Lin, S.; Chen, G.; Ouyang, G. Application of the NU-1000 Coated SPME Fiber on Analysis of Trace Organochlorine Pesticides in Water. *Anal. Chim. Acta* **2022**, *1218*, 339982. [CrossRef]

66. Junior, J.; Repoppi, N. Determination of Organochlorine Pesticides in Ground Water Samples Using Solid-Phase Microextraction by Gas Chromatography-Electron Capture Detection. *Talanta* **2007**, *72*, 1833–1841. [CrossRef] [PubMed]

67. Li, H.-P.; Li, G.-C.; Jen, J.-F. Determination of Organochlorine Pesticides in Water Using Microwave Assisted Headspace Solid-Phase Microextraction and Gas Chromatography. *J. Chromatogr. A* **2003**, *1012*, 129–137. [CrossRef]

68. Perez-Trujillo, J.P.; Frías, S.; Sáchez, M.J.; Conde, J.E.; Rodríguez-Delgado, M.A. Determination of Organochlorine Pesticides by Gas Chromatography with Solid-Phase Microextraction. *Chromatographia* **2002**, *56*, 191–197. [CrossRef]

69. Abolghasemi, M.M.; Hassani, S.; Bamorowat, M. Efficient Solid-Phase Microextraction of Triazole Pesticides from Natural Water Samples Using a Nafion-Loaded Trimethylsilane-Modified Mesoporous Silica Coating of Type SBA-15. *Microchim. Acta* **2016**, *183*, 889–895. [CrossRef]

70. Derouiche, A.; Driss, M.R.; Morizur, J.P.; Taphanel, M.H. Simultaneous Analysis of Polychlorinated Biphenyls and Organochlorine Pesticides in Water by Headspace Solid-Phase Microextraction with Gas Chromatography-Tandem Mass Spectrometry. *J. Chromatogr. A* **2007**, *1138*, 231–243. [CrossRef] [PubMed]

71. Domínguez, I.; Arrebola, F.J.; Gavara, R.; Martínez Vidal, J.L.; Frenich, A.G. Automated and Simultaneous Determination of Priority Substances and Polychlorinated Biphenyls in Wastewater Using Headspace Solid Phase Microextraction and High Resolution Mass Spectrometry. *Anal. Chim. Acta* **2017**, *1002*, 39–49. [CrossRef]

72. Chang, S.M.; Doong, R.A. Concentration and Fate of Persistent Organochlorine Pesticides in Estuarine Sediments Using Headspace Solid-Phase Microextraction. *Chemosphere* **2006**, *62*, 1869–1878. [CrossRef]

73. López, F.J.; Pitarch, E.; Egea, S.; Beltran, J.; Hernández, F. Gas Chromatographic Determination of Organochlorine and Organophosphorus Pesticides in Human Fluids Using Solid Phase Microextraction. *Anal. Chim. Acta* **2001**, *433*, 217–226. [CrossRef]

74. Tsoukali, H.; Theodoridis, G.; Raikos, N.; Grigoratou, I. Solid Phase Microextraction Gas Chromatographic Analysis of Organophosphorus Pesticides in Biological Samples. *J. Chromatogr. B* **2005**, *822*, 194–200. [CrossRef]

75. Hernandez, F.; Pitarch, E.; Beltran, J.; Lopez, F.J. Headspace Solid-Phase Microextraction in Combination with Gas Chromatography and Tandem Mass Spectrometry for the Determination of Organochlorine and Organophosphorus Pesticides in Whole Human Blood Q. *J. Chromatogr. B* **2002**, *769*, 65–77. [CrossRef]

76. Zhu, F.; Ruan, W.; He, M.; Zeng, F.; Luan, T.; Tong, Y.; Lu, T.; Ouyang, G. Application of Solid-Phase Microextraction for the Determination of Organophosphorus Pesticides in Textiles by Gas Chromatography with Mass Spectrometry. *Anal. Chim. Acta* **2009**, *650*, 202–206. [CrossRef] [PubMed]

77. Leong, W.-H.; Teh, S.-Y.; Hossain, M.M.; Nadarajaw, T.; Zabidi-Hussin, Z.; Chin, S.-Y.; Lai, K.-S.; Lim, S.-H.E. Application, Monitoring and Adverse Effects in Pesticide Use: The Importance of Reinforcement of Good Agricultural Practices (GAPs). *J. Environ. Manag.* **2020**, *260*, 109987. [CrossRef] [PubMed]

78. European Commission. *Regulation (EC) No 396/2005, Maximum Residue Levels of Pesticides in or on Food and Feed of Plant and Animal Origin*; Publications Office of the European Union: Copenhagen, Denmark, 2005.

79. Page, B.D.; Lacroix, G. Application of Solid-Phase Microextraction to the Headspace Gas Chromatographic Analysis of Halogenated Volatiles in Selected Foods. *J. Chromatogr. A* **1993**, *648*, 199–211. [CrossRef]

80. Mascrez, S.; Psillakis, E.; Purcaro, G. A Multifaceted Investigation on the Effect of Vacuum on the Headspace Solid-Phase Microextraction of Extra-Virgin Olive Oil. *Anal. Chim. Acta* **2020**, *1103*, 106–114. [CrossRef]

81. Möder, M.; Popp, P.; Eisert, R.; Pawliszyn, J. Determination of Polar Pesticides in Soil by Solid Phase Microextraction Coupled to High-Performance Liquid Chromatography-Electrospray/Mass Spectrometry. *Fresenius. J. Anal. Chem.* **1999**, *363*, 680–685. [CrossRef]

82. Hernandez, F.; Beltran, J.; Lopez, F.J.; Gaspar, J.V. Use of Solid-Phase Microextraction for the Quantitative Determination of Herbicides in Soil and Water Samples. *Anal. Chem.* **2000**, *72*, 2313–2322. [CrossRef]

83. European Commission. *2002/657/EC: Commission Decision of 12 August 2002 Implementing Council Directive 96/23/EC Concerning the Performance of Analytical Methods and the Interpretation of Results (Text with EEA Relevance) (Notified under Document Number C(2002) 3044)*; Publications Office of the European Union: Copenhagen, Denmark, 2002.

84. Nollet, L.M.L.; De Gelder, L.S.P. (Eds.) *Handbook of Water Analysis*; CRC Press: Boca Raton, FL, USA, 2000; ISBN 9780849384868.

85. Barceló, D. Environmental Protection Agency and Other Methods for the Determination of Priority Pesticides and Their Transformation Products in Water. *J. Chromatogr.* **1993**, *643*, 117–143. [CrossRef]

86. Ghiasvand, A.; Yazdankhah, F.; Paull, B. Heating-, Cooling- and Vacuum-Assisted Solid-Phase Microextraction (HCV-SPME) for Efficient Sampling of Environmental Pollutants in Complex Matrices. *Chromatographia* **2020**, *83*, 531–540. [CrossRef]

87. Psillakis, E. Vacuum-Assisted Headspace Solid-Phase Microextraction: A Tutorial Review. *Anal. Chim. Acta* **2017**, *986*, 12–24. [CrossRef]

88. Lv, F.; Gan, N.; Huang, J.; Hu, F.; Cao, Y.; Zhou, Y.; Dong, Y.; Zhang, L.; Jiang, S. A Poly-Dopamine Based Metal-Organic Framework Coating of the Type PDA-MIL-53(Fe) for Ultrasound-Assisted Solid-Phase Microextraction of Polychlorinated Biphenyls Prior to Their Determination by GC-MS. *Microchim. Acta* **2017**, *184*, 2561–2568. [CrossRef]

89. Wei, M.C.; Jen, J.F. Determination of Polycyclic Aromatic Hydrocarbons in Aqueous Samples by Microwave Assisted Headspace Solid-Phase Microextraction and Gas Chromatography/Flame Ionization Detection. *Talanta* **2007**, *72*, 1269–1274. [CrossRef]
90. Mascrez, S.; Purcaro, G. Exploring Multiple-cumulative Trapping Solid-phase Microextraction for Olive Oil Aroma Profiling. *J. Sep. Sci.* **2020**, *43*, 1934–1941. [CrossRef]
91. Spadafora, N.D.; Mascrez, S.; McGregor, L.; Purcaro, G. Exploring Multiple-Cumulative Trapping Solid-Phase Microextraction Coupled to Gas Chromatography–Mass Spectrometry for Quality and Authenticity Assessment of Olive Oil. *Food Chem.* **2022**, *383*, 132438. [CrossRef]

*Review*

# New Prospects in the Electroanalysis of Heavy Metal Ions (Cd, Pb, Zn, Cu): Development and Application of Novel Electrode Surfaces

Vasiliki Keramari, Sophia Karastogianni and Stella Girousi *

Analytical Chemistry Laboratory, School of Chemistry, Faculty of Sciences, 54124 Thessaloniki, Greece; vasilikik20keramari@gmail.com (V.K.); skarastogianni@hotmail.com (S.K.)
* Correspondence: girousi@chem.auth.gr

**Abstract:** The detection of toxic heavy metal ions, especially cadmium (Cd), lead (Pb), zinc (Zn), and copper (Cu), is a global problem due to ongoing pollution incidents and continuous anthropogenic and industrial activities. Therefore, it is important to develop effective detection techniques to determine the levels of pollution from heavy metal ions in various media. Electrochemical techniques, more specifically voltammetry, due to its properties, is a promising method for the simultaneous detection of heavy metal ions. This review examines the current trends related to electrode formation and analysis techniques used. In addition, there is a reference to advanced detection methods based on the nanoparticles that have been developed so far, as well as formation with bismuth and the emerging technique of screen-printed electrodes. Finally, the advantages of using these methods are highlighted, while a discussion is presented on the benefits arising from nanotechnology, as it gives researchers new ideas for integrating these technologies into devices that can be used anywhere at any time. Reference is also made to the speciation of metals and how it affects their toxicity, as it is an important subject of research.

**Keywords:** heavy metal ions; modified electrode; electroanalysis; nanoparticles

**Citation:** Keramari, V.; Karastogianni, S.; Girousi, S. New Prospects in the Electroanalysis of Heavy Metal Ions (Cd, Pb, Zn, Cu): Development and Application of Novel Electrode Surfaces. *Methods Protoc.* **2023**, *6*, 60. https://doi.org/10.3390/mps6040060

Academic Editor: Verónica Pino

Received: 10 May 2023
Revised: 10 June 2023
Accepted: 21 June 2023
Published: 26 June 2023

**Copyright:** © 2023 by the authors. Licensee MDPI, Basel, Switzerland. This article is an open access article distributed under the terms and conditions of the Creative Commons Attribution (CC BY) license (https://creativecommons.org/licenses/by/4.0/).

## 1. Introduction

Most heavy metal ions exist naturally in the environment; however, some come from anthropogenic sources, such as some industries, agriculture, the burning of fossil fuels, insecticides, car exhausts, and sewage. These heavy metal ions in large quantities can become hazardous to the biological system. For example, cadmium (Cd), lead (Pb), zinc (Zn), and copper (Cu) affect the environment due to their non-biodegradability and accumulated toxicity [1].

In the ground, all of the inorganic elements that are necessary and essential for the normal growth and development of plants exist. Even though some heavy metal ions, such as copper (Cu) and zinc (Zn) etc., are necessary for various enzymatic functions, an excessive concentration of heavy metal ions can cause serious problems [2–4], as they can become toxic and dangerous, with serious environmental implications. Toxic heavy metal ions vary in their nature and mode of accumulation, either in the soil or in plants. Some of the sources of heavy metal ions in the soil are fertilizers, pesticides, and sewage sludge [5].

Toxic metals such as cadmium (Cd) and lead (Pb), as well as many others, can easily end up in the higher members of the biological food chain and, therefore, in humans, causing serious diseases such as gastrointestinal tract (GIT) infections, cardiovascular problems, bone problems, and even cancer [6,7]. On the other hand, equally serious are the effects that heavy metal ions have on the environment, e.g., soil pollution, which is one of the most important problems for the planet. The term "soil pollution" refers to the concentration of polluting substances in soil in quantities that cause a change in the composition of the soil, resulting in disturbances in the ecosystem.

In order to limit the negative effects that heavy metals ion have on both humans and the environment, it is necessary to accurately determine the concentrations of heavy metal ions in their sources of accumulation.

Over the years, various techniques have been established for the detection of heavy metal ions (HMIs), including inductively coupled plasma mass spectrometry (ICP-MS) [8], inductively coupled plasma optical emission spectrometry (ICP-OES) [9], inductively coupled plasma atomic emission spectrometry (ICP-AES) [10], flame atomic absorption spectrophotometry (FAAS) [11], and atomic absorption spectroscopy (AAS) [12], where sometimes the emphasis is on the parameters and sometimes on the choice of the analysis method. Although they provide high accuracy and sensitivity, spectrometric methods such as atomic absorption (AAS) and inductively coupled plasma mass spectrometry (ICP-MS) are accompanied by certain limitations, such as high costs and the fact that they are time-consuming and do not allow on-site measurements.

However, as already mentioned, researchers' interest in recent years has focused on the identification of heavy metal ions with the help of various electrochemical methods, particularly voltammetry, as it is an easy, fast, and relatively inexpensive way to determine them compared to other analytical methods. Although previous techniques are very sensitive and selective, due to the limitations they cause, electrochemical methods such as voltammetry are preferred for the detection of heavy metal ions, which, in contrast to previous techniques, have the advantages of low cost, simplicity, ease of operation, fast analysis, portability, the ability to monitor environmental samples in the field, and high sensitivity and selectivity [13]. Electrochemical techniques, especially voltammetry, involve electroanalytical methods for the determination of one or more analytes by measuring the current as a function of potential. However, voltammetry is the only electrochemical method that has high sensitivity and can be applied for the on-site recognition and detection of heavy metal ions [14]. Voltammetric modifications of ESA (i.e., DPASV, SWASV, and AdSV), including two variants of stripping potentiometry (PSA and CCSA), have historically been widely recognized as powerful techniques for detecting heavy metal ions due to their remarkable sensitivity, which allows for the detection at trace and ultra-trace levels.

Furthermore, there is also electrochemical-stripping analysis (ESA).

In order to improve sensitivity and selectivity in the detection of heavy metal ions and to examine their toxicity on both the environment and human health, we combine electrochemical techniques with certain modifiers that facilitate the detection process. Due to their characteristic properties, electrode modifiers were used in combination with electroanalytical methods.

The need for alternative electrode modifiers rose as attention began to turn more and more towards green chemistry, which aims to reduce the use and production of hazardous substances such as mercury [15], and since certain European regulations prohibit the export and storage of metallic mercury. Since then, several metals have been tested for their ability to replace mercury (e.g., bismuth). In addition, different organic and inorganic membranes have been evaluated for their potential application in detecting Pb with ESA.

In recent years, nanoparticles (NPs) have emerged as a promising area of research, replacing various electrode-shaping media, thanks to their unique physicochemical properties that differ from their bulk counterparts. NPs are defined as particles with a dimension of at least 100 nm, and are generally classified into four categories, including metal and metal oxide, semiconductors, polymers, and lipids. They exhibit a high surface area to volume ratio, which enhances their reactivity and allows them to interact with biological systems and the environment in unique ways. The ability of NPs to cross biological membranes and barriers has attracted considerable attention in various fields, including medicine, environmental science, and engineering, and they have been widely used for heavy metal detection, such as that for Cd, Pb, Zn, and Cu, due to their unique optical, electrical, and magnetic properties.

This review mainly discusses the use of voltammetry in the simultaneous detection of the presence of two or more heavy metal ions in different media using modified electrodes and presents a comprehensive overview of the modifiers for various electrodes. This paper includes a historical review from the abolition of mercury to its replacement and the discovery of innovative nanoparticles and presents their applications in chemistry and the environment. The present review also aims to summarize the different types of nanoparticles, such as metallic, semiconducting, and carbon-based nanoparticles, and their application in various electroanalytical techniques, including voltammetry.

## 2. Electrodes

Initially, the measurement technique is selected, with the most used measurement techniques for detecting heavy metal ions being SW, SWASV, and DPASW. In the next step, the appropriate working electrode is selected, where carbonaceous electrodes are dominant. They appear to further improve the performance of the voltammetric methods, as they are flexible, offer a wide potential window, and have desirable conductive and surface properties that allow for the sensitive determination of analytes. The four most common are glassy carbon electrodes (GCE), graphite electrodes (GE), carbon paste electrodes (CPE), and screen-printed carbon electrodes (SPCE). These electrodes are widely used for the determination of heavy metal ion concentrations ($Cd^{2+}$, $Pb^{2+}$, and $Zn^{2+}$), while some of the most common types of modified electrodes include, among others, nanoparticle-modified electrodes, chemically modified electrodes (using chemical modifiers such as bismuth (Bi) and, formerly, mercury (Hg)), carbon-based modified electrodes, and enzyme-modified electrodes.

Therefore, the appropriate electrode modifier is further investigated. In our case, we work with GCE, and our working electrode is modified with bismuth. In a recent study for the determination of lead, which is a heavy metal ion, and with GCE as the working electrode, modification was performed with BFS (blast furnace slag), which is an economically efficient process and a new material in the field of detection, with many promising results [16]. A typical electrochemical analytical system consists mainly of the following three parts: an electrochemical detection device, an electrochemical detection instrument, and an electrolyte. The electrochemical detection instrument usually consists of the following three electrodes: a working electrode (WE), a reference electrode (RE), and a counter electrode (CE). After modifying the surfaces of the Wes using different materials, they can be used for the specific detection of various types of metallic ions. A representative illustration of sample preparation is quoted schematically in Figure 1, where the main steps followed are shown. In some Wes, surface modification is necessary because the nature of the used electrode can greatly affect the sensitivity and the selectivity of the analytical procedure. For example, it is essential to polish the surface of a glassy carbon electrode (GCE) with a polishing cloth posing 0.1 mm and 0.005 mm alumina powder, inducing a mirror-like surface, which improves the analytical features of the detection procedure [14].

### 2.1. Glassy Carbon Electrode for HM Detection

One of the most common electrodes used for the detection and quantitative determination of heavy metals is glassy carbon electrodes (GCEs), which are produced by pyrolysis of polymeric resins. Some of their distinguishing characteristics are that they are easily based, non-reactive, resistant to high temperatures, and impermeable to gases and liquids, while simultaneously providing excellent analytical performance across a wide range of metals. With GCE, relatively low limits of detection (LOD) can be achieved [17], as demonstrated, for example, by Thanh et al., who modified the electrode surface with a Bi film and achieved LOD values of 1.07, 0.93, 0.65, and 0.94 ppb for Zn, Cd, Pb, and Cu, respectively, while maintaining high accuracy and repeatability within the measurements [18]. In another study, Hassan et al. determined Pb in the linear range of 10.0–120.0 $\mu g\ L^{-1}$, Cd and Zn in the linear range of 0.0–50 $\mu g\ L^{-1}$ (corresponding LOD values: 3.18 ng $L^{-1}$, 0.107 $\mu g\ L^{-1}$, and 0.037 $\mu g\ L^{-1}$) in tap water simultaneously using SWASV with Hg–Bi film electrodeposition on GCE in the presence of poly(1,2-diaminoanthraquinone) [19].

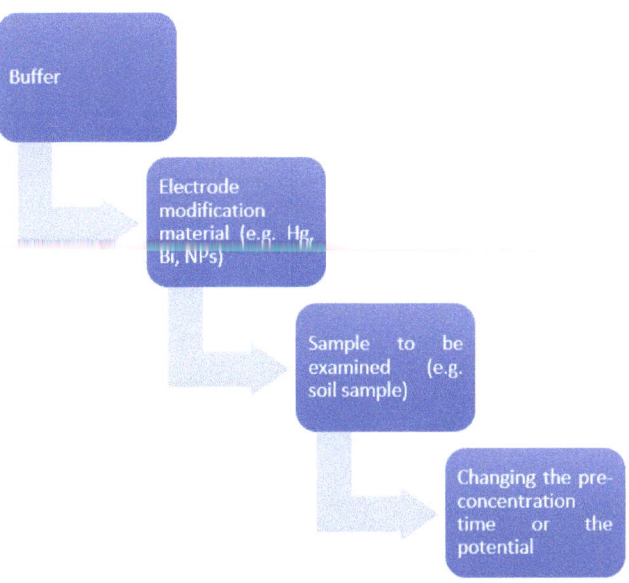

**Figure 1.** Sample preparation procedures for measurement.

*2.2. Graphite Electrode for HM Detection*

Graphite electrodes (GEs) are an effective tool in the electroanalytical determination of heavy metals, mainly preferred for the detection of Cd, Pb, and Cu. Their large surface areas, conductive properties, and low cost are characteristics that make GE an ideal choice for the electrochemical analysis of HMs [17]. For example, Donmez et al. were able to simultaneously determine Pb and Cd in water samples up to 0.46 and 1.11 µg L$^{-1}$, respectively [20].

*2.3. Carbon Paste Electrode for HM Detection*

For the electroanalysis of heavy metals, carbon paste electrodes (CPEs) are commonly used due to the simplicity of preparation and formatting. To prepare a CPE, the carbonaceous material in the form of powder or powder in a high-viscosity oil, usually paraffin, mineral oil, or pyrite oil, needs to be added. They are often preferred because it is relatively easy to incorporate various substances, such as nanoparticles, waste materials, and various chemicals, into this mixture in order to enhance metal deposition on the surface [17]. In a study conducted by Zheng et al., the simultaneous detection of Cd and Pb was performed using CPE formatted with mesoporous alumina, yielding respective LOD results of 0.2 and 2.0 nM, respectivley [21]. However, with the appropriate formatters, suitable LOD levels can be achieved using CPE, at either lower or higher levels.

*2.4. Screen-Printed Electrode (SPE) for HM Detection*

Over the past two decades, significant effort has been made towards the development of more environmentally friendly and "green" electrode materials as substitutes, primarily for mercury electrodes. Screen-printed electrodes (SPEs), which are cost-effective electrochemical substrates, have undergone significant improvements in recent decades in terms of their form and printing materials. The main advantages of using SPEs are their flexibility and the fact that they can be used as disposable sensors (avoiding any potential contamination from previous measurements). Additionally, compared to other analytical methods, they are economically efficient, easily tunable, and suitable for integration into portable devices. SPEs may be the most suitable electrochemical sensors for in situ analyses due to their linear performance, low power requirements, fast response, high sensitivity, and ability to operate at room temperature [22,23].

SPEs are chosen as an economical substrate for electrochemical biosensor applications due to their small size, ease of mass production, and portability. To create an SPE, the electrodes must be prepared as pastes or inks so that they cannot be printed on the predetermined surface called the substrate, which can be made from ceramic or polymer material such as polyethylene terephthalate (PET) [17]. The surface of the SPE can be easily modified to suit multiple purposes related to different pollutants. There are many materials that can be used as the modifiers of SPEs for environmental analysis, including noble metals, inorganic nanoparticles, enzymes, and DNA sequences. For the determination of heavy metals such as Cd, Pb, Cu, and Zn, they can be modified with bismuth either in situ [24] or ex situ [25]. Bismuth is characterized as the most common modifier of SPEs for electroanalysis due to its good analytical performance and "environmentally friendly" characteristics [26]. Alternatively, bismuth can be deposited on SPEs in various forms, such as bismuth oxide [27–29], bismuth nanoparticles [30], and pre-deposited bismuth as a film [31]. Another material for the modification of SPEs is gold (Au), which is a valuable tool for the determination of Cu. Some researchers modified chitosan (CTS) on the surface of SPCEs for the simultaneous determination of Pb (II), Cu(II), Cd(II), and Hg(II), with a preconcentration time of only 30 s [32]. However, this specific material presents challenges in the determination of other heavy metals, as its cathodic overpotential for the reduction in hydrogen ions is low and the range of its cathodic polarization is limited. Therefore, Au electrodes are not particularly useful for the detection of metals with a more cathodic oxidative potential, such as Cd, Pb, and Zn [33]. They can also be modified with nanoparticles, such as carbon nanoparticles, for the determination of $Cd^{2+}$, $Pb^{2+}$, and $Cu^{2+}$ ions [34,35], or chemically synthesized bismuth metallic nanoparticles that modify the screen-printed carbon electrodes that are used for the detection of Zn(II), Cd(II), and Pb(II) in liquid samples [30]. For example, McEleney et al., for the determination of Cd and Zn, modified the surface of their electrode, which was made of graphene oxide and graphitic carbon nitride, with bismuth and gallium by changing the pH within the same cell [35].

## 3. Modifiers

### 3.1. Mercury (Hg)-Based Electrode Variants and Bismuth Electrode Modifiers for HM Detection

To investigate the toxicity caused by heavy metal ions in soil, atmosphere, and consequently human health, and with the goal of limiting it, both techniques of analytical chemistry, specifically voltammetry in our case, and "supporting" means such as mercury (Hg) and bismuth (Bi), have been used, while in recent years nanoparticles (NPs) have been used.

For many years, mercury was used as the material for the modification of working electrodes used in trace element detection due to its high sensitivity, reproducibility, and renewability. Mercury-based electrodes have been widely used for several decades in the detection of heavy metal ions using electrochemical techniques, thanks to their large cathodic window, reproducibility, sensitivity, and low background [36–38].

However, mercury is a heavy metal that has become increasingly unpopular for use, due to its intense toxicity and bioaccumulation in many species [38–42]. The danger associated with mercury-based electrodes is in their use, handling, and disposal due to their toxicity. In addition, it has been repeatedly shown that the absorption of Hg harms human health, as it can lead to many serious problems, such as neurological consequences (as it penetrates the blood–brain barrier), memory loss, insomnia, neuromuscular changes, and various effects on the renal system.

Over the years, various materials, such as noble metals (Pt, Pd, Au, and Ag) and other metals (Ru, Cu, Co, Ni, Pb, Sb, Bi, and Al), have been proposed and tested to replace mercury in the electrode modification process [43,44]. Although it is a heavy metal ion, the metal that prevailed is bismuth, due to its low toxicity [38,39,41,45,46], as well as its similar electroanalytical properties to mercury, such as a wide potential window, simple preparation, partial insensitivity to dissolved oxygen, and ability to form alloys with different metals [38–40,47,48]. Bismuth is also environmentally friendly [49] and has mostly succeeded in replacing mercury, as the latter is quite toxic. Therefore, around 2000, electrodes modified with bismuth were introduced that were constructed from a layer of bismuth deposited on a suitable substrate [39,50] and represented a very attractive alternative solution to the commonly used mercury electrodes [45]. Many different materials have been used as electrode substrates, such as carbon, glassy carbon, carbon fibers, carbon paste, graphite, wax-impregnated graphite, gold, and platinum [33–40,50]. The current peaks obtained in the voltammograms when using bismuth electrodes tend to be sharp and well-defined [45], allowing for the reliable, fast, and economical recognition and quantification of metals present in the sample. Due to its characteristics, bismuth can be used as a film in electrodes, such as in glassy carbon electrodes (GCEs), and then can be used in various sample analyses (environmental, biological, etc.).

An example worth mentioning is the simultaneous detection of different heavy metals ions that are present in a sample, which is performed following the modification of vitreous carbon electrode with bismuth. Then various experimental parameters, such as potential and deposition time, are optimized, and finally, the appropriate voltammetric method, square wave voltammetry (SWV), is used [14].

Electrochemical detection focuses on developing new electrode materials with better properties compared to commercial electrodes. The performance of the voltammetric determination of heavy metal ions depends heavily on the properties of the working electrode. Working electrodes can be modified with different materials to allow for specific recognitions and concentrations of metal ions. Additionally, it has been reported that the deposition of metal membranes on nanocarbon materials can further improve the electrochemically active surface [15,51]. Among these, bismuth (Bi) film not only has low toxicity, high sensitivity, and a strong response signal, but it can also form binary or multiple-component alloys with heavy metal ions, which is a process that is analogical to the amalgamation with mercury, also enhancing the efficiency of the deposition at the surface (of either elemental mercury or bismuth).

One of the earliest applications of a bismuth-modified electrode was for the determination of lead in water samples using electrochemical stripping analysis (ESA), and, because it is considered one of the least toxic metals, it has subsequently been used for analyses in the medical and pharmaceutical sectors [52]. For approximately 20 years, bismuth-modified electrodes, which emerged as a replacement for toxic mercury, have found a wide range of environmental and clinical applications. Therefore, bismuth films are often combined with carbon materials for cooperative heavy metal detection. Hutton et al. [53] used a bismuth film for stripping measurements of cobalt and cadmium internal soil extracts. Recently, Bi-modified electrodes have also been successfully used in the electrochemical detection of nitrophenols, while bismuth oxides have been used in the detection of paracetamol.

*3.2. Inorganic Materials as Electrode Modifiers for HM Detection*

Another method for the detection of HMs is the surface modification of electrodes with inorganic materials, as this method can improve the sensitivity, stability, and selectivity of the electrode for HM ion detection. It has been found that inorganic nanoparticles modified on the electrode surface can adsorb more HM ions, thereby enhancing the specific surface area of the working electrode. They can also play a catalytic role in the deposition of HM ions on the electrode surface, thereby improving the electrochemical detection capability. However, a disadvantage of this is that inorganic nanoparticles are relatively expensive and challenging to produce on a large scale [54].

Some of the inorganic materials that have been successfully used for electrode modification and HM detection are metal and metal oxide nanoparticles, such as noble metal nanoparticles (e.g., AuNPs), bimetallic, and metal oxide nanoparticles. They have been employed to modify the electrode surface due to their favorable optical and electrical properties. They can be combined with other chemicals and biomolecules to construct various highly specialized electrochemical detection devices for HM ions. An example of this is the electrodeposition of AuNPs and Bi film on a screen-printed carbon electrode (SPCE) to obtain Bi/AuNP/SPCE, where the synergistic effect of the Bi membrane and AuNPs increased the surface area of the electrode, with good electrical conductivity. Using the differential-pulse anodic stripping voltammetry (DPASV) method, with detection limits of 50 ng/L ($Zn^{2+}$), 20 ng/L ($Pb^{2+}$), and 30 ng/L ($Cu^{2+}$), the successful simultaneous detection of $Zn^{2+}$, $Pb^{2+}$, and $Cu^{2+}$ in lake water was achieved [55].

### 3.3. Nanoparticles as Electrode Modifiers for HM Detection

As we have already mentioned, pollution from heavy metal ions is a significant issue, and, currently, the addition of NPs with electrochemical sensors has developed a significant and innovative analytical technique for the detection of heavy metal ions (HMs), as nanomaterials have been shown to offer remarkable properties as detection platforms. Nanomaterials could be considered as a promising tool for the scientific community to detect toxic heavy metal ions, due to their sensitivity and selectivity, fast response time, high sensitivity, and reproducibility, as well as the possibility of the simultaneous detection of HMs with very low detection and quantification limits [56]. Over time, many different modification techniques have been explored. Recent studies have shown that NP-modified electrodes can be very useful in electrochemical sensor technology if they are designed and constructed correctly [57]. Their surface area-to-volume ratio is high, and, in combination with the characteristics exhibited by NPs, such as those based on metals and metal oxides, polymers, and carbon, they can be beneficial for removing HMs from the environment [58].

Nanotechnology and nanoparticles (NPs) have transformed science and technology. Today, this field has advanced to such a degree that it allows the development of the production of nanoparticles using various physical, chemical, and even biological techniques. Among these techniques, the one that stands out and is preferred more in the industrial sector to produce nanoparticles is the biological method, due to its ease, the need for mild operating conditions, and the production of more environmentally friendly products and waste [59]. Most industries today exploit the chemical properties of nanoparticles, as they are unique compared to their counterparts in volume, which is determined by their size, shape, composition, and surface chemistry and can be adapted to various applications. Some of the most important chemical properties of nanoparticles are as follows [60]:

- The high surface-to-volume ratio: NPs have a high surface-to-volume ratio, which makes them extremely reactive. This property can be used in various applications, such as catalysis and sensors.
- Surface energy: The surface energy of NPs is high due to the presence of unsaturated surface atoms. This property affects the agglomeration, stability, and dispersion of NPs.
- Electromagnetic properties: NPs can exhibit unique electromagnetic properties due to their size, shape, and composition. For example, gold NPs exhibit localized surface plasmon resonance, which can be used for sensing and imaging applications.
- Surface chemistry: The surface chemistry of NPs can be tailored by modifying their surface functional groups, which can change their surface reactivity and chemical properties.
- Oxidation-reduction properties: NPs can exhibit unique oxidation-reduction properties, due to their small size and large surface area. This can be utilized in various applications, including energy storage and conversion.

The synthesis of NPs using the bio reduction method has drawn scientific interest, as it has managed to overcome the drawbacks of using conventional chemical methods, such as thermodynamic stability, monodispersity, and particle formation [61]. The biogenic synthesis of NPs presents some advantages over chemical synthesis, such as the absence of the need for high temperatures, toxic chemicals, pressure, energy, radiation processes, laser ablation, and ultraviolet and ultrasonic fields, as well as the fact that the biomolecules required for NP synthesis are abundant and easily accessible, such as the availability in marine sources [62]. On the other hand, NPs produced from the noble metal group, such as gold (Au) and silver (Ag), exhibit interesting chemical and electromagnetic properties, such as chemical stability, conductivity, and good optical properties, due to their ability to interact with electromagnetic radiation [63,64].

NPs, due to their large surface area, are excellent electron mediators. Therefore, NPs suitable electrode surface modifier the improvement of the analytical characteristics of electrodes. For instance, silicon (Si)- and carbon-based nanoparticles and have been successfully used as electrode modifiers. By using this kind of modification, the behavior of these NPs improves, and the constructed electrochemical sensors can measure the analytes in nanoscale. Thus, the use of NPs as electrode surface modifiers increases the active surface area, catalytic activity, conductivity and makes the response of the used electrodes more rapid. These redesigned sensors can also exhibit size-dependent characteristics and can have better functional units [65]. Currently, the addition of these NPs to electrochemical sensors has developed a significant analytical technique for detecting heavy metal ions (HMs).

Gold is excellent for the fabrication of nanomaterials because gold nanoparticles (AuNPs) are characterized as excellent templates for the development of cutting-edge chemical and biological sensors, thanks to their unique physical and chemical properties. AuNPs can be easily produced and made very stable [66]. They also have exceptional optical-electronic properties, and, with the right linkers, they offer a high surface-to-volume ratio and great biocompatibility. Furthermore, AuNPs can provide a versatile substrate for attaching a wide variety of chemical or biological moieties, allowing the selective capture and detection of small molecules and biological targets. It must be stressed that, when HMs are analyzed, particularly mercury and lead, different materials are incorporated with AuNPs, while the same materials can also be used for the detection of Cd (II) and Pb(II) [67]. According to the composition conditions, gold nanoparticles (AuNPs) appear in a variety of shapes, such as spherical, which is the most common shape used in the electrochemical detection of heavy metals, with sizes ranging from 4 to 298 nm. Different composition shapes of AuNPs were tested for the detection of Pb (II). According to the literature, for Pb (II) detection, Dutta et al. synthesized nano-stars, which were prepared by mixing an auric chloride solution with 4-(2-hydroxyethyl)-l-piperazineethanesulfonic acid (HEPES) without stirring or agitation, and boiling the nano-stars for 5 min resulted in spherical nanoparticles. The same process was later used for the synthesis of gold nano-stars for Cd (II) detection [68]. To evaluate the concentrations of cadmium in different water sources (such as lake, sewage, tap water, and groundwater), a glassy carbon electrode was modified with AuNPs, l-cysteine, and reduced graphene oxide, and, by applying square-wave voltammetry, the best performance for Cd (II) detection was achieved. The same electrode also exhibited the highest reported sensitivity for Pb (II) detection [69].

Other NPs, such as superparamagnetic $Fe_3O_4$@EDTA, have been developed for the simultaneous adsorption and removal of Zn(II), Pb(II), and Cd(II) from different environmental water and soil samples. For this method, which has been proven to be simple, fast, effective, sensitive with high removal yields, reproducible, and repeatable, electrodes modified with polymeric EDTA were used for the detection of various metallic ions at different pH values [70]. Furthermore, after the adsorption process, easy separation is provided only by the application of an external magnetic field. In conclusion, this method is an effective and less time-consuming technique for the simultaneous adsorption and removal of heavy metal ion targets in different environmental water and soil samples [71]. For the synthesis of $Fe_3O_4$@EDTA nanoparticles, the following procedure was performed: 15 mmol of $FeCl_3 \cdot 6H_2O$ and 7.5 mmol of $FeCl_2 \cdot 4H_2O$ were dissolved in 150 mL of deionized water under a nitrogen atmosphere at room temperature with vigorous stirring. Then, 50 mL of 25% $NH_4OH$ solution was added to the stirring mixture under intense mechanical stirring, adjusting the pH to 11, while simultaneously adding EDTA solution (3 mmol in 30 mL of water). The resulting black dispersion was continuously stirred for 1 h at room temperature and then refluxed for 2 h [71]. Finally, the resulting nanoparticles were isolated using a magnetic field; washed with water, ethanol, and diethyl ether; and dried. The simultaneous adsorption and removal of Zn(II), Pb(II), and Cd(II) in different environmental samples were successfully achieved with the aid of the superparamagnetic nano-adsorbent $Fe_3O_4$@EDTA.

Another category of NPs, AgNPs, which are used as electrode modifiers for the detection of heavy metal ions such as Cd(II) and Cu(II), have received significant attention due to some characteristics they exhibit, such as good electrical conductivity, high specific area, and an easy synthesis method [72,73]. It is supported by literature that, when the electrochemical technique is combined with nanomaterials, a very fast and efficient detection of heavy metal ions is obtained.

For the simultaneous determination of lead and cadmium, $MnCo_2O_4$ nanoparticles have been successfully used, which were morphed on a glassy carbon electrode. $MnCo_2O_4$ nanoparticles exhibit exceptional electrochemical properties, such as a fast current response, a low detection limit, and good selectivity, due to their unique structure [74]. The synthesis of $MnCo_2O_4$ nanoparticles was carried out using the citric gel combustion method as follows: solutions of manganese nitrate and cobalt nitrate were mixed in a molar ratio of 1:2, and citric acid was used as the fuel. The stoichiometric ratio of citric acid, according to the existing literature for the nitrate groups, was 1:3.6 [69], and the pH was adjusted to seven by adding an ammonium hydroxide solution. The mixture was then heated to approximately 80 °C in an open glass beaker under continuous stirring conditions (100 rpm) until a light-pink colloidal solution was formed, which was transformed into a gel and finally calcined at 450 °C for 2 h, resulting in the formation of black $MnCo_2O_4$ nanoparticles. For the measurement of Pb(II) and Cd(II) content in water samples, a glassy carbon electrode modified with $MnCo_2O_4$NPs was used. The electrode was immersed in a supporting electrolyte solution of $H_2SO_4$/KCl containing Pb(II) and Cd(II) during the pre-concentration stage. In this stage, the accumulation and reduction of metal ions to metal ($M^{2+}$ to $M^0$) occurred, while, in the deposition stage, the opposite process took place, i.e., the re-oxidation of metals ($M^0$) and the stripping of metal ion species ($M^{2+}$) from the solution. The electrochemical response was measured using the electrochemical technique of linear sweep anodic stripping voltammetry (LSASV) in the potential range of −1.0 to 0 V. The glassy carbon–$MnCo_2O_4$NPs electrode exhibited excellent electrochemical properties, such as a fast current response, a low detection limit, and good selectivity, due to its unique structure, as well as a satisfactory detection performance for Cd(II) and Cu(II) [74].

Lee et al. used tin nanoparticles (SnNPs) with reduced graphene oxide on a glassy carbon electrode to determine Cd(II), Pb(II), and Cu(II) [75]. For the simultaneous detection of Cd(II), Pb(II), and Cu(II) ions, G-Sn/GCS electrodes were used, which were derived when reduced graphene oxide (RGO) was activated with tin nanoparticles (SnNPs) and cast onto glassy carbon sheets (GCS), followed by electrochemical reduction. The results showed that the G-Sn/GCS electrodes exhibited good stability, high sensitivity, and good repeatability in heavy metal detection.

Bismuth nanoparticles (BiNPs) have attracted interest as pre-concentrators for the detection of heavy metals such as cadmium and lead ions, while they are also used as working electrode modifiers in stripping electrochemical analysis. Among the various reported methods for the synthesis of BiNPs, we have focused on the typical polyol method, which is widely used for these types of metallic and semimetallic nanoparticles [76]. Several techniques based on bismuth sulfide in combination with different working electrodes have been tested for the detection of heavy metals. Using square-wave anodic stripping voltammetry and a bismuth-nanodust-modified electrode, Lee et al. succeeded in detecting zinc, cadmium, and lead ions, followed by the preparation of spherical bismuth with different particle size distributions in order to investigate their effect on the sensitivity and detection limits of the detected metals. It was observed that, as the particle size decreased (from 406 to 166 nm), both the sensitivity and the detection limit improved [77]. On the other hand, Rico et al. [34] used the method of Lee et al. to detect heavy metals by forming a printed carbon electrode. The optimization of the method included accumulation configuration, resulting in better detection limits in the flow cells for Zn(II), Cd(II), and Pb(II) than in the batch cells. In another study for the determination of mercury, cadmium, lead, and copper ions using differential-pulse anodic stripping voltammetry, Sahoo et al. formed a graphene-oxide- and bismuth-nanoparticle-modified carbon paste electrode, with particle sizes ranging from 40 to 100 nm. A glassy carbon electrode was modified with a micro-nanoparticle/bismuth membrane by Saturno et al. for the determination of cadmium and lead using differential-pulse voltammetry. However, in almost all of these cases, the problem of copper (II) interference at high concentrations arose, which was addressed most of the time [67].

The application of palladium nanoparticles (Pd NPs) for the detection of heavy metals has been tested by a few research groups. However, they all synthesized porous activated carbon (PAC), followed by decorating PAC with palladium nanoparticles using a single-step thermal reduction method (with slightly different conditions). Zhang et al. used spherical Pd NPs, with a size range of 20–30 nm, for the individual and simultaneous determination of Cd(II), Pb(II), and Cu(II) through SWASV (square-wave anodic stripping voltammetry). The obtained detection limits for Cd(II), Pb(II), and Cu(II) were found to be lower in the individual determinations compared to the simultaneous determinations. The technique was successfully tested in practical water samples, although the nature of the water was not specified [78]. Summary of NPs-assisted detection of heavy metals ions is shown in Table 1.

**Table 1.** Summary of NP-assisted detection of heavy metal ions.

| Nanoparticles (NPs) | HM Detection | Ref. |
|---|---|---|
| AuNPs | Cd(II) and Pb(II) | [67,79] |
| $Fe_3O_4$@EDTA-NPs | Cd(II), Pb(II), and Zn(II) | [70,80,81] |
| AgNPs | Cd(II) and Cu(II) | [72,73] |
| $MnCo_2O_4$NPs | Cd(II) and Pb(II) | [74] |
| SnNPs | Cd(II), Pb(II), and Cu(II) | [75] |
| BiNPs | Cd(II) and Pb(II) | [76] |
| PdNPs | Cd(II), Pb(II), and Cu(II) | [67,78] |

As can be seen, in addition to their electrocatalytic properties, these nanomaterial-based electrodes have the advantages of low cost, high sensitivity, and convenient functionality, making them highly promising for practical applications in heavy metal detection. However, further research is required in order to overcome potential issues and improve the stability and selectivity of these sensors.

*3.4. Ion-Imprinted Polymers as Electrode Modifiers for HM Detection*

The determination of heavy metals in water and food intended for human consumption has led to an alternative method based on the modification of a working electrode with ion-imprinted polymers (IIPs). By using IIPs immobilized on a carbon paste electrode (CPE), the determination of both cadmium and lead ions can be achieved. The base of an IIP-modified CPE (CPEs-IIP) is usually formed by modifying a binary mixture, to which ingredients such as imprinted polymers are added or incorporated. The quantity can range from 10% to 30% of the composite mass, allowing for several recognition sites on the electrode surface, which correlates with the current intensity received. Additionally, IIPs can also be immobilized on a glassy carbon electrode (GCE) surface to detect cadmium, lead, cadmium, and pseudo silver ions in drinking water and food samples. In recent years, there has been significant interest in the CPEs-IIP technique, as it represents a simple and cost-effective method for the detection and analysis of heavy metal ions ($Cd^{2+}$ and $Pb^{2+}$) in both drinking water and food. This makes it a highly promising technique for improving everyday life [82].

*3.5. Speciation of HMs*

The speciation of chemical heavy metals is an important factor that alters the toxicity of heavy metals. The potential mobility, bioavailability, and environmental behavior of heavy metals depend largely on their specific chemical forms and existing conditions. Depending on the existing environmental conditions, various types of metal can exist as metals and metalloids, which can be present as hydroxides, organometallic compounds, biomolecules, and other forms, such as inorganic ions in the form of cations (e.g., Cd(II), Pb(II)), or anions (e.g., As(III) and As(V)). The determination of these molecular species is called metal speciation. Considering the toxicity and the bioavailability of heavy metals, their speciation is often more significant than determining total HMs. However, there are few publications that address the analysis and determination of different forms of a specific heavy metal, despite there being significant progress in the development of fluorescence detectors for detecting the total concentration of HMs [76].

Most of the current notification analysis methods for HMs combine separation techniques, such as gas chromatography, high-performance liquid chromatography, and so on, as well as detection techniques such as atomic absorption spectroscopy, atomic emission spectroscopy, and inductively coupled plasma mass spectrometry [83]. Although these methods have many advantages, a significant limitation is their requirements for a series of complex pre-processing steps, which are time-consuming and laborious. Electrochemical techniques have been characterized as the easiest, fastest, and most economical for species analysis. Van den Berg implemented the speciation analysis of different metallic elements such as iron, molybdenum, copper, and so on, using electrochemical methods [84–86]. ASV has been successfully adopted as the most common method for the analysis of metallic species, especially for unstable species [83]. Although voltammetric methods yield excellent results for the analysis of the speciation of many metal ions, there are certain issues and limitations that need to be addressed promptly, such as the relationship between bioavailability and metal speciation, which remains ambiguous and requires further study [87].

One example where speciation plays a crucial role is the exposure of plants to heavy metals through the aqueous phase of the soil, the soil solution, which contains heavy metals in various forms, such as free metal ions, simple inorganic complexes, and complexes with organic ligands. The composition of the soil solution can be influenced by the properties of the plants, as well as the soil itself. Environmental monitoring and the speciation of heavy metals (HMs) in soil solutions are highly important for ecological assessments and for understanding the plant–soil relationship [88]. Equally important is understanding the relationship between bioavailability and trace element speciation in natural aquatic environments [87].

## 4. Conclusions

Given that the global environmental burden of heavy metal ions and the associated impact on health and the environment are increasing, the interest in improving the quality of life and reducing their effects is ongoing. This review provides a general discussion on the field of electrochemical detection of HMs using bismuth-modified electrodes, which have replaced toxic mercury-modified electrodes, as well as nanomaterials as a more modern form of electrode modification, and their use in voltammetric experiments.

Certain materials, such as bismuth, have been distinguished for their ease of use, which is why researchers prefer them. The selection of suitable electrode modification materials is very important, as they improve the electrochemical properties of the electrode, increase its effective surface area for the transfer of the electrochemical signal, and produce detectable signals that are suitable for the indirect detection of HMIs. For the detection of HMs, voltammetric methods have been distinguished as the most powerful and sensitive and least time consuming.

In conclusion, the various nanoparticles that have been tested for the detection of heavy metal ions have shown significant results, due to the advantage of their large surface area compared to their size, as well as their electrocatalytic properties.

The purpose of the review was also to draw the attention of researchers working in electrochemistry, to develop new, improved morphology-controlled electrodes for the simultaneous detection of HMs at very low permissible limits (ppm, ppb), and, thus, reduce the quantity and toxicological burden of HMs in the environment [89].

**Author Contributions:** Conceptualization, V.K. and S.G.; methodology, V.K., S.G. and S.K.; software, V.K. and S.K.; validation, V.K., S.G. and S.K.; formal analysis, V.K. and S.K.; investigation, V.K. and S.K.; resources, V.K., S.G. and S.K.; data curation, V.K. and S.K.; writing—original draft preparation, V.K. and S.K.; writing—review and editing, V.K., S.G. and S.K.; visualization, V.K. and S.K.; supervision, S.G. and S.K.; project administration, V.K., S.G. and S.K.; funding acquisition, V.K., S.G. and S.K. All authors have read and agreed to the published version of the manuscript.

**Funding:** This research received no external funding.

**Institutional Review Board Statement:** Not applicable.

**Informed Consent Statement:** Not applicable.

**Data Availability Statement:** Not applicable.

**Conflicts of Interest:** The authors declare no conflict of interest.

**Abbreviations**

HMs, Heavy metal ions; ICP-MS, Inductively coupled plasma mass spectrometry; FAAS, Flame atomic absorption spectrometry; AAS, Atomic absorption spectrometry; ICP-OES, Inductively coupled plasma optical emission spectroscopy; ICP-AES, Inductively coupled plasma atomic emission spectroscopy; LSV, Linear sweep voltammetry; WE, Working electrode; ESA, Electrochemical-stripping analysis; RE, Reference electrode; CE, Counter electrode; EDTA, Ethylenediamine tetraacetic acid; SPEs, Screen-printed electrodes; CTS, Chitosan; AgNPs, Silver nanoparticle; AuNPs, Gold nanoparticles; NPs, Nanoparticles; SWASV, Square-wave anodic stripping voltammetry; GE, Graphite electrode; CPE, Carbon paste electrode; SPCE, Screen-printed carbon electrode; BFs, Blast furnace slag; Hg, Mercury; Bi, Bismuth; Cd, Cadmium; Pb, Lead; Zn, Zinc; Cu, Copper; Si, Silica; CV, Cyclic voltammetry; DPV, Different pulse voltammetry; GCE, Glassy carbon electrode; SWV, Square-wave voltammetry.

# References

1. Aragay, G.; Pons, J.; Merkoçi, A. Recent trends in macro-, micro-, and nanomaterial-based tools and strategies for heavy-metal detection. *Chem. Rev.* **2011**, *111*, 3433–3458. [CrossRef]
2. Wang, C.; Li, W.; Guo, M.; Ji, J. Ecological risk assessment on heavy metals ions in soils: Use of soil diffuse reflectance mid-infrared Fourier-transform spectroscopy. *Sci. Rep.* **2017**, *7*, 40709. [CrossRef] [PubMed]
3. Li, M.; Gou, H.; Al-Ogaidi, I.; Wu, N. Nanostructured sensors for detection of heavy metals ions: A review. *ACS Sustain. Chem. Eng.* **2013**, *1*, 713–723. [CrossRef]
4. Guascito, M.R.; Malitesta, C.; Mazzotta, E.; Turco, A. Inhibitive determination of metal ions by an amperometric glucose oxidase biosensor: Study of the effect of hydrogen peroxide decomposition. *Sens. Actuators B Chem.* **2008**, *131*, 394–402. [CrossRef]
5. Alengebawy, A.; Abdelkhalek, S.T.; Qureshi, S.R.; Wang, M.Q. Heavy metals ions and Pesticides Toxicity in Agricultural Soil and Plants: Ecological Risks and Human Health Implications. *Toxics* **2021**, *9*, 42. [CrossRef]
6. Xu, M.; Hadi, P.; Chen, G.; McKay, G. Removal of cadmium ions from wastewater using innovative electronic waste-derived material. *J. Hazard. Mater.* **2014**, *273*, 118–123. [CrossRef]
7. Yu, S.; Pang, H.; Huang, S.; Tang, H.; Wang, S.; Qiu, M.; Chen, Z.; Yang, H.; Song, G.; Fu, D. Recent advances in metal-organic framework membranes for water treatment: A review. *Sci. Total Environ.* **2021**, *800*, 149662. [CrossRef]
8. Koelmel, J.; Amarasiriwardena, D. Imaging of metal bioaccumulation in Hay-scented fern (*Dennstaedtia punctilobula*) rhizomes growing on contaminated soils by laser ablation ICP-MS. *Environ. Pollut.* **2012**, *168*, 62–70. [CrossRef]
9. Massadeh, A.M.; Alomary, A.A.; Mir, S.; Momani, F.A.; Haddad, H.I.; Hadad, Y.A. Analysis of Zn, Cd, As, Cu, Pb, and Fe in snails as bioindicators and soil samples near traffic road by ICP-OES. *Environ. Sci. Pollut. Res.* **2016**, *23*, 13424–13431. [CrossRef]
10. Rao, K.S.; Balaji, T.; Rao, T.P.; Babu, Y.; Naidu, G.R.K. Determination of iron, cobalt, nickel, manganese, zinc, copper, cadmium and lead in human hair by inductively coupled plasma-atomic emission spectrometry. *Spectrochim. Acta Part B At. Spectrosc.* **2002**, *57*, 1333–1338.
11. Daşbaşı, T.; Saçmacı, Ş.; Çankaya, N.; Soykan, C. A new synthesis, characterization and application chelating resin for determination of some trace metals in honey samples by FAAS. *Food Chem.* **2016**, *203*, 283–291. [CrossRef] [PubMed]

12. Siraj, K.; Kitte, S.A. Analysis of copper, zinc and lead using atomic absorption spectrophotometer in ground water of Jimma town of Southwestern Ethiopia. *Int. J. Chem. Anal. Sci.* **2013**, *4*, 201–204. [CrossRef]
13. Mei, C.J.; Ahmad, S.A.A. A review on the determination heavy metals ions using calixarene-based electrochemical sensors. *Arab. J. Chem.* **2021**, *14*, 103303.
14. Lu, Y.; Liang, X.; Niyungeko, C.; Zhou, J.; Xu, J.; Tian, G. A review of the identification and detection of heavy metal ions in the environment by voltammetry. *Talanta* **2018**, *178*, 324–338. [CrossRef] [PubMed]
15. Armenta, S.; Garrigues, S.; de la Guardia, M. Green Analytical Chemistry. *Trends Anal. Chem.* **2008**, *27*, 497. [CrossRef]
16. Mourya, A.; Mazumdar, B.; Sinha, S.K. Determination and quantification of heavy metal ion be electrochemical method. *J. Environ. Chem. Eng.* **2019**, *7*, 103459. [CrossRef]
17. Ustabasi, G.S.; Ozcan, M.; Yilmaz, I. Review-Voltammetric Determination of Heavy Metals with Carbon-Based Electrodes. *J. Electrochem. Soc.* **2021**, *168*, 097508. [CrossRef]
18. Thanh, N.M.; Van Hop, N.; Luyen, N.D.; Phong, N.H.; Toan, T.T.T. Simultaneous determination of Zn(II), Cd(II), Pb(II), and Cu(II) using differential pulse anodic stripping voltammetry at a bismuth film-modified electrode. *Adv. Mater. Sci. Eng.* **2019**, *2019*, 1826148. [CrossRef]
19. Hassan, K.M.; Gaber, S.E.; Altahan, M.F.; Azzem, M.A. Single and simultaneous voltammetric sensing of lead(II), cadmium(II) and zinc(II) using a bimetallic Hg-Bi supported on poly(1,2-diaminoanthraquinone)/glassy carbon modified electrode. *Sens. Bio Sens. Res.* **2020**, *29*, 100369. [CrossRef]
20. Dönmez, K.B.; Çetinkaya, E.; Deveci, S.; Karadağ, S.; Şahin, Y.; Doğu, M. Preparation of electrochemically treated nanoporous pencil-graphite electrodes for the simultaneous determination of Pb and Cd in water samples. *Anal. Bioanal. Chem.* **2017**, *409*, 4827–4837. [CrossRef]
21. Zheng, X.; Chen, S.; Chen, J.; Guo, Y.; Peng, J.; Zhou, X.; Lv, R.; Lin, J.; Lin, R. Highly sensitive determination of lead(II) and cadmium(II) by a large surface area mesoporous alumina modified carbon paste electrode. *RSC Adv.* **2018**, *8*, 7883. [CrossRef]
22. Tse, Y.-H.; Janda, P.; Lam, H.; Lever, A.B.P. Electrode with electropolymerized tetraaminophthalocyanatocobalt(II) for detection of sulfide ion. *Anal. Chem.* **1995**, *67*, 981–985. [CrossRef]
23. Hori, Y.; Takahashi, R.; Yoshinami, Y.; Murata, A. Electrochemical reduction of CO at a copper electrode. *Phys. Chem. B* **1997**, *101*, 7075–7081. [CrossRef]
24. Noh, M.F.M.; Tothill, I.E. Determination of lead (II), cadmium (II) and copper (II) in waste-water and soil extracts on mercury film screen-printed carbon electrodes sensor. *Sains Malays* **2011**, *40*, 1153–1163.
25. Parat, C.; Aguilar, D.; Authier, L.; Potin-Gautier, M.; Companys, E.; Puy, J.; Galceran, J. Determination of Free Metal Ion Concentrations Using Screen-Printed Electrodes and AGNES with the Charge as Response Function. *Electroanalysis* **2011**, *23*, 619–627. [CrossRef]
26. Li, M.; Li, Y.-T.; Li, D.-W.; Long, Y.-T. Recent developments and applications of screen-printed electrodes in environmental assays—A review. *Anal. Chim. Acta* **2012**, *734*, 31–44. [CrossRef]
27. Hwang, G.-H.; Han, W.-K.; Park, J.-S.; Kang, S.-G. An electrochemical sensor based on the reduction of screen-printed bismuth oxide for the determination of trace lead and cadmium. *Sens. Actuators B Chem.* **2008**, *135*, 309–316. [CrossRef]
28. Kadara, R.O.; Tothill, I.E. Development of disposable bulk-modified screen-printed electrode based on bismuth oxide for stripping chronopotentiometric analysis of lead (II) and cadmium (II) in soil and water samples. *Anal. Chim. Acta* **2008**, *623*, 76–81. [CrossRef]
29. Kadara, R.O.; Jenkinson, N.; Banks, C.E. Disposable bismuth oxide screen printed electrodes for the high throughput screening of heavy metals ions. *Electroanalysis* **2009**, *21*, 2410–2414.
30. Rico, M.G.; Olivares-Marín, M.; Gil, E.P. Modification of carbon screen-printed electrodes by adsorption of chemically synthesized Bi nanoparticles for the voltammetric stripping detection of Zn(II), Cd(II) and Pb(II). *Talanta* **2009**, *80*, 631–635. [CrossRef]
31. Serrano, N.; Díaz-Cruz, J.M.; Ariño, C.; Esteban, M. Ex situ Deposited Bismuth Film on Screen-Printed Carbon Electrode: A Disposable Device for Stripping Voltammetry of Heavy Metal Ions. *Electroanalysis* **2010**, *22*, 1460–1467. [CrossRef]
32. Khaled, E.; Hassan, H.N.A.; Habib, I.H.I.; Metelka, R. Chitosan modified screen-printed carbon electrode for sensitive analysis of heavy metals ions. *Int. J. Electrochem. Sci.* **2010**, *5*, 158–167.
33. Economou, A. Screen-Printed Electrodes Modified with "Green" Metals for Electrochemical Stripping Analysis of Toxic Elements. *Sensors* **2018**, *18*, 1032. [CrossRef]
34. Aragay, G.; Pons, J.; Merkoçi, A. Enhanced electrochemical detection of heavy metals ions at heated graphite nanoparticle-based screen-printed electrodes. *J. Mater. Chem.* **2011**, *21*, 4326–4331. [CrossRef]
35. Mc Eleney, C.; Alves, S.; Mc Crudden, D. Novel determination of Cd and Zn in soil extract by sequential application of bismuth and gallium thin films at a modified screen-printed carbon electrode. *Anal. Chim. Acta* **2020**, *1137*, 94. [CrossRef] [PubMed]
36. Domingos, R.F.; Huidobro, C.; Companys, E.; Galceran, J.; Puy, J.; Pinheiro, J. Comparison of AGNES (absence of gradients and Nernstian equilibrium stripping) and SSCP (scanned stripping chronopotentiometry) for trace metal speciation analysis. *J. Electroanal. Chem.* **2008**, *617*, 141–148. [CrossRef]
37. Aguilar, D.; Galceran, J.; Companys, E.; Puy, J.; Parat, C.; Authier, L.; Potin-Gautier, M. Non-purged voltammetry explored with AGNES. *Phys. Chem. Chem. Phys.* **2013**, *15*, 17510. [CrossRef]
38. Cao, L.; Jia, J. Sensitive determination of Cd and Pb by differential pulse stripping voltammetry with in situ bismuth-modified zeolite doped carbon paste electrodes. *Electrochim. Acta* **2008**, *53*, 2177. [CrossRef]

39. Kokkinos, C.; Raptis, I. A Economou and T Speliotis. Disposable micro-fabricated electrochemical bismuth sensors for the determination of Tl(I) by stripping voltammetry. *Procedia Chem.* **2009**, *1*, 1039.
40. Economou, A. Bismuth-film electrodes: Recent developments and potentialities for electroanalysis. *Trends Anal. Chem.* **2005**, *24*, 334.
41. Kokkinos, C.; Economou, A. Novel disposable bismuth-sputtered electrodes for the determination of trace metals by stripping voltammetry. *Electrochem. Commun.* **2007**, *9*, 2795.
42. Hutton, E.; Ogorevc, B.; Hočevar, S.; Weldon, F.; Smyth, M.R.; Wang, J. An introduction to bismuth film electrode for use in cathodic electrochemical detection. *Electrochem. Commun.* **2001**, *3*, 707.
43. Arduini, F.; Quintana, J. Bismuth-modified electrodes for lead detection. *Trends Anal. Chem.* **2010**, *2–11*, 1295.
44. Pauliukaite, R.; Hočevar, S.; Ogorevc, B.; Wang, J. Characterization and Applications of a Bismuth Bulk Electrode. *Electroanalysis* **2004**, *16*, 719. [CrossRef]
45. Wang, J. Stripping Analysis at Bismuth Electrodes: A Review. *Electroanalysis* **2005**, *17*, 1341–1346. [CrossRef]
46. Xu, H.; Zeng, L. A Nafion-coated bismuth film electrode for the determination of heavy metals in vegetable using differential pulse anodic stripping voltammetry: An alternative to mercury-based electrodes. *Food Chem.* **2008**, *109*, 834. [CrossRef] [PubMed]
47. Hočevar, S.; Švancara, I.; Vytřas, K.; Ogorevc, B. Novel electrode for electrochemical stripping analysis based on carbon paste modified with bismuth powder. *Electrochem. Acta* **2005**, *51*, 706.
48. Baldrianova, L.; Švancara, I.; Vlcek, M.; Economou, A.; Sotiropoulos, S. Effect of Bi(III) concentration on the stripping voltammetric response of in situ bismuth-coated carbon paste and gold electrodes. *Electrochim. Acta* **2006**, *52*, 481. [CrossRef]
49. Sam, K. *The Disappearing Spoon (and Other True Tales of Madness, Love, and the History of the World from the Periodic Table of Elements)*; Back Bay Books: New York, NY, USA; Boston, MA, USA, 2011; pp. 158–160.
50. Kokinos, C.; Economous, A. Lithographically fabricated disposable bismuth-film electrodes for the trace determination of Pb(II) and Cd(II) by anodic stripping voltammetry. *Electrochim. Acta* **2008**, *53*, 294. [CrossRef]
51. Yıldız, C.; Eskiköy Bayraktepe, D.; Yazan, Z. Highly sensitive direct simultaneous determination of zinc(II), cadmium(II), lead(II), and copper(II) based on in-situ-bismuth and mercury thin-film plated screen-printed carbon electrode. *Mon. Chem. Chem. Mon.* **2021**, *152*, 1527–1537. [CrossRef]
52. Naumov, A.V. World market of bismuth: A review. *Russ. J. Non-Ferr. Met.* **2007**, *48*, 10. [CrossRef]
53. Hutton, E.A.; van Elteren, J.T.; Ogorevc, B.; Smyth, M.R. Validation of bismuth film electrode for determination of cobalt and cadmium in soil extracts using ICP-MS. *Talanta* **2004**, *63*, 849. [CrossRef] [PubMed]
54. Wu, Q.; Bi, H.-M.; Han, X.-J. Research Progress of Electrochemical Detection of Heavy Metal Ions. *Chin. J. Anal. Chem.* **2021**, *49*, 330–340. [CrossRef]
55. Lu, Z.; Zhang, J.; Dai, W.; Lin, X.; Ye, J.; Ye, J. A screen-printed carbon electrode modified with a bismuth film and gold nanoparticles for simultaneous stripping voltammetric determination of Zn(II), Pb(II) and Cu(II). *Microchim. Acta* **2017**, *184*, 4731–4740. [CrossRef]
56. Okpara, E.C.; Fayemi, O.E.; Wojuola, O.B.; Onwudiwe, D.C.; Ebenso, E.E. Electrochemical detection of selected heavy metals in water: A case study of African experiences. *RSC Adv.* **2022**, *12*, 26319–26361. [CrossRef]
57. Hassan, M.H.; Khan, R.; Andreescu, S. Advances in Electrochemical Detection Methods for Measuring Contaminants of Emerging Concerns. *Electrochem. Sci. Adv.* **2021**, *2*, e2100184. [CrossRef]
58. Ali, Z.; Ullah, R.; Tuzen, M.; Ullah, S.; Rahim, A.; Saleh, T.A. Colorimetric sensing of heavy metals ions on metal doped metal oxide nanocomposites: A review. *Trends Environ. Anal. Chem.* **2023**, *37*, e00187. [CrossRef]
59. Ibraheem, I.B.M.; Abd-Elaziz, B.E.E.; Saad, W.F.; Fathy, W.A. Green biosynthesis of silver nanoparticles using marine Red Algae Acanthophora specifera and its antimicrobial activity. *J. Nanomed. Nanotechnol.* **2016**, *7*, 409.
60. Joudeh, N.; Linke, D. Nanoparticle classification, physicochemical properties, characterization, and applications: A comprehensive review for biologists. *J. Nanobiotechnol.* **2022**, *20*, 262. [CrossRef]
61. Li, S.; Niu, Y.; Chen, H.; He, P. Complete genome sequence of an Arctic Ocean bacterium *Shewanella* sp. Arc9-LZ with capacity of synthesizing silver nanoparticles in darkness. *Mar. Genom.* **2021**, *56*, 100808. [CrossRef]
62. Abdel-Raoof, A.M.; El-Shal, M.A.; Said, R.A.M.; Abostate, M.H.; Morshedy, S.; Emara, M.S. Versatile sensor modified with gold nanoparticles carbon paste electrode for anodic stripping determination of brexpiprazole: A voltammetric study. *J. Electrochem. Soc.* **2019**, *166*, B948. [CrossRef]
63. Ahmed, M.B.; Zhou, J.L.; Ngo, H.H.; Guo, W.; Chen, M. Progress in the preparation and application of modified biochar for improved contaminant removal from water and wastewater. *Bioresour. Technol.* **2016**, *214*, 836–851. [CrossRef]
64. Panahi, A.; Levendis, Y.A.; Vorobiev, N.; Schiemann, M. Direct observations on the combustion characteristics of Miscanthus and Beechwood biomass including fusion and spheroidization. *Fuel Process. Technol.* **2017**, *166*, 41–49. [CrossRef]
65. Huang, W.; Zhang, Y.; Li, Y.; Zeng, T.; Wan, Q.; Yang, N. Morphology-controlled electrochemical sensing of environmental $Cd^{(2+)}$ and $Pb^{(2+)}$ ions on expanded graphite supported $CeO_2$ nanomaterials. *Anal. Chim. Acta* **2020**, *1126*, 63–71. [CrossRef] [PubMed]
66. Adeniji, T.M.; Stine, K.J. Nanostructure Modified Electrodes for Electrochemical Detection of Contaminants of Emerging Concern. *Coatings* **2023**, *13*, 381. [CrossRef]
67. Sawan, S.; Maalouf, R.; Errachid, A.; Jaffrezic-Renault, N. Metal and metal oxide nanoparticles in the voltammetric detection of heavy metals ions: A review. *Trends Anal. Chem.* **2020**, *131*, 116014. [CrossRef]

68. Lu, D.; Sullivan, C.; Brack, E.M.; Drew, C.P.; Kurup, P. Simultaneous voltammetric detection of cadmium(II), arsenic(III), and selenium(IV) using gold nanostar–modified screen-printed carbon electrodes and modified Britton-Robinson buffer. *Anal. Bioanal. Chem. Anal. Bioanal. Chem.* **2020**, *412*, 4113–4125. [CrossRef]
69. Jelić, D.; Zeljković, S.; Škundrić, B.; Mentus, S. Thermogravimetric study of the reduction of CuO–WO$_3$ oxide mixtures in the entire range of molar ratios. *J. Therm. Anal. Calorim.* **2018**, *132*, 77–90. [CrossRef]
70. Rahman, A.; Park, D.S.; Won, M.-S.; Park, S.-M.; Shim, Y.-B. Selective electrochemical analysis of various metal ions at an EDTA bonded conducting polymer modified electrode. *Electroanalysis* **2004**, *16*, 1366–1370. [CrossRef]
71. Ghasemi, E.; Heydari, A.; Sillanpää, M. Superparamagnetic Fe$_3$O$_4$@EDTA nanoparticles as an efficient adsorbent for simultaneous removal of Ag(I), Hg(II), Mn(II), Zn(II), Pb(II) and Cd(II) from water and soil environmental samples. *Microchem. J.* **2017**, *131*, 51–56. [CrossRef]
72. Hassan, K.M.; Elhaddad, G.M.; AbdelAzzem, M. Voltammetric determination of cadmium(II), lead(II) and copper(II) with a glassy carbon electrode modified with silver nanoparticles deposited on poly(1,8-diaminonaphthalene). *Microchim. Acta* **2019**, *186*, 440. [CrossRef] [PubMed]
73. Zhang, K.; Zhang, N.; Zhang, L.; Xu, J.; Wang, H.; Wang, C.; Geng, T. Amperometric sensing of hydrogen peroxide using a glassy cabon electrode modified with silver nanoparticles on poly(alizarin yellow R). *Microchim. Acta* **2011**, *173*, 135–141. [CrossRef]
74. Antunović, V.; Ilić, M.; Baošić, R.; Jelić, D.; Lolić, A. Synthesis of MnCo$_2$O$_4$ nanoparticles as modifiers for simultaneous determination of Pb(II) and Cd(II). *Public Libr. Sci.* **2019**, *14*, e0210904. [CrossRef] [PubMed]
75. Lee, P.M.; Chen, Z.; Li, L.; Liu, E. Reduced graphene oxide decorated with tin nanoparticles through electrodeposition for simultaneous determination of trace heavy metals ions. *Electrochim. Acta* **2015**, *174*, 207–214. [CrossRef]
76. Cadevall, M.; Ros, J.; Merkoc, A. Bismuth nanoparticles integrationinto heavy metal electrochemical strippingsensor. *Electrophoresis* **2015**, *36*, 1872–1879. [CrossRef] [PubMed]
77. Lee, G.J.; Kim, C.K.; Lee, M.K.; Rhee, C.K. Simultaneous voltammetric determination of Zn, Cd and Pb at bismuth nanopowder electrodes with various particle size distributions. *Electroanalysis* **2010**, *22*, 530–535. [CrossRef]
78. Zhang, T.; Jin, H.; Fang, Y.; Guan, J.; Ma, S.; Pan, Y.; Zhang, M.; Zhu, H.; Liu, X.; Du, M. Detection of trace $Cd^{2+}$, $Pb^{2+}$ and $Cu^{2+}$ ions via porous activated carbon supported palladium nanoparticles modified electrodes using SWASV. *Mater. Chem. Phys.* **2019**, *225*, 433–442. [CrossRef]
79. Lahari, S.A.; Amreen, K.; Dubey, S.K.; Ponnalagu, R.N.; Goel, S. Optimized porous carbon-fibre microelectrode for multiplexed, highly reproducible and repeatable detection of heavy metals ions in real water samples. *Environ. Res.* **2023**, *220*, 115192. [CrossRef]
80. Üstündağ, Z.; Solak, A.O. EDTA modified glassy carbon electrode: Preparation and characterization. *Electrochim. Acta* **2009**, *54*, 6426–6432. [CrossRef]
81. Rahman, M.A.; Won, M.S.; Shim, Y.B. Characterization of an EDTA bonded conducting polymer modified electrode: Its application for the simultaneous determination of heavy metal ions. *Anal. Chem.* **2003**, *75*, 1123–1129. [CrossRef]
82. Rebolledo-Perales, L.E.; Romero, G.A.; Ibarra-Ortega, I.S.; Galán-Vidal, C.A.; Pérez-Silva, I. Review-Electrochemical Determination of Heavy Metals in Food and Drinking Water Using Electrodes Modified with Ion-Imprinted Polymers. *J. Electrochem. Soc.* **2021**, *168*, 067516. [CrossRef]
83. Han, H.; Pan, D.; Zhang, S.; Wang, C.; Hu, X.; Wang, Y.; Pan, F. Simultaneous Speciation Analysis of Trace Heavy metals ions (Cu, Pb, Cd and Zn) in Seawater from Sishili Bay, North Yellow Sea, China. *Bull. Environ. Contam. Toxicol.* **2018**, *101*, 486–493. [CrossRef]
84. Bi, Z.; Salaün, P.; Berg, C.M.v.D. The speciation of lead in seawater by pseudopolarography using a vibrating silver amalgam microwire electrode. *Mar. Chem.* **2013**, *151*, 1–12. [CrossRef]
85. van den Berg, C.M. Chemical speciation of iron in seawater by cathodic stripping voltammetry with dihydroxynaphthalene. *Anal. Chem.* **2006**, *78*, 156–163. [PubMed]
86. Whitby, H.; van den Berg, C.M.G. Evidence for copper-binding humic substances in seawater. *Mar. Chem.* **2015**, *173*, 282–290. [CrossRef]
87. Han, H.; Pan, D. Voltammetric methods for speciation analysis of trace metals in natural waters. *Trends Environ. Anal. Chem.* **2021**, *29*, e00119. [CrossRef]
88. Dytrtová, J.J.; Šestáková, I.; Jakl, M.; Száková, J.; Miholová, D.; Tlustoš, P. The use of differential pulse anodic stripping voltammetry and diffusive gradient in thin films for heavy metals ions speciation in soil solution. *Cent. Eur. J. Chem.* **2008**, *6*, 71–79.
89. Rassaei, L.; Marken, F.; Sillanpää, M.; Amiri, M.; Cirtiu, C.M.; Sillanpää, M. Nanoparticles in electrochemical sensors for environmental monitoring. *Trends Anal. Chem.* **2011**, *30*, 1704–1715. [CrossRef]

**Disclaimer/Publisher's Note:** The statements, opinions and data contained in all publications are solely those of the individual author(s) and contributor(s) and not of MDPI and/or the editor(s). MDPI and/or the editor(s) disclaim responsibility for any injury to people or property resulting from any ideas, methods, instructions or products referred to in the content.

Article

# Simultaneous Quantification of Bisphenol-A and 4-Tert-Octylphenol in the Live Aquaculture Feed *Artemia franciscana* and in Its Culture Medium Using HPLC-DAD

Despoina Giamaki [1], Konstantina Dindini [1], Victoria F. Samanidou [2] and Maria Touraki [1,*]

1. Laboratory of General Biology, Division of Genetics, Development and Molecular Biology, Department of Biology, School of Sciences, Aristotle University of Thessaloniki (A.U.TH.), 54 124 Thessaloniki, Greece; despina_giamaki@imbb.forth.gr (D.G.); konstantina.dindini@stud.ki.se (K.D.)
2. Laboratory of Analytical Chemistry, Department of Chemistry, School of Sciences, Aristotle University of Thessaloniki (A.U.TH.), 54 124 Thessaloniki, Greece; samanidu@chem.auth.gr
* Correspondence: touraki@bio.auth.gr; Tel.: +30-2310-99-82-92

**Citation:** Giamaki, D.; Dindini, K.; Samanidou, V.F.; Touraki, M. Simultaneous Quantification of Bisphenol-A and 4-Tert-Octylphenol in the Live Aquaculture Feed *Artemia franciscana* and in Its Culture Medium Using HPLC-DAD. *Methods Protoc.* **2022**, *5*, 38. https://doi.org/10.3390/mps5030038

Academic Editor: Joselito P. Quirino

Received: 4 April 2022
Accepted: 29 April 2022
Published: 1 May 2022

**Publisher's Note:** MDPI stays neutral with regard to jurisdictional claims in published maps and institutional affiliations.

**Copyright:** © 2022 by the authors. Licensee MDPI, Basel, Switzerland. This article is an open access article distributed under the terms and conditions of the Creative Commons Attribution (CC BY) license (https://creativecommons.org/licenses/by/4.0/).

**Abstract:** Aquaculture, a mass supplier of seafood, relies on plastic materials that may contain the endocrine disruptors bisphenol-A (BPA) and tert-octylphenol (t-OCT). These pollutants present toxicity to *Artemia*, the live aquaculture feed, and are transferred through it to the larval stages of the cultured organisms. The purpose of this work is the development and validation of an analytical method to determine BPA and t-OCT in *Artemia* and their culture medium, using n-octylphenol as the internal standard. Extraction of the samples was performed with $H_2O$/TFA (0.08%)–methanol (3:1), followed by SPE. Analysis was performed in a Nucleosil column with mobile phases A (95:5, $v/v$, 0.1% TFA in $H_2O$:$CH_3CN$) and B (5:95, $v/v$, 0.08% TFA in $H_2O$:$CH_3CN$). Calibration curves were constructed in the range of concentrations expected following a 24 h administration of BPA (10 µg/mL) or t-OCT (0.5 µg/mL), below their respective $LC_{50}$. At the end of exposure to the pollutants, their total levels appeared reduced by about 32% for BPA and 35% for t-OCT, and this reduction could not be accounted for by photodegradation (9–19%). The developed method was validated in terms of linearity, accuracy, and precision, demonstrating the uptake of BPA and t-OCT in *Artemia*.

**Keywords:** bisphenol A; 4-tert-octylphenol; *Artemia franciscana*; HPLC-DAD

## 1. Introduction

The worldwide accumulation of plastic pollution in the marine environment poses a great risk towards the wellbeing of the vertebrate and invertebrate organisms that live in it, as well as of humans as the end consumer. The protective gear that was extensively produced during the COVID-19 pandemic, has greatly contributed to the plastic waste, a large amount of which ends up in the marine environment, where leaching occurs of its hydrophobic additives [1]. The highly hydrophobic organic contaminants include endocrine disrupting chemicals (EDCs), such as bisphenol A (BPA), nonylphenol, and tert-octylphenol (t-OCT). EDCs are incorporated into plastics as building blocks or stabilizers [2] and leach out, due to aging and heat [3,4]. Bisphenol-A (4, 4′-Isopropylidenediphenol) is widely used as a component of synthetic plastic [2] while 4-tert-octylphenol (4-(1,1,3,3-tetramethylbutyl) phenol) is used as a component of polyethoxylates, applied in detergents, industrial cleaners, and emulsifiers [3]. The percentage of stabilizers and antioxidants, including BPA and t-OCT in plastics, depends on the chemical structure of the produced plastic polymer, ranging from 0.05 to 3% $w/w$ [5]. Studies on the levels of BPA and t-OCT in microplastics are gaining attention, and marine microplastics sampled in the open Pacific and the Atlantic Ocean, as well as from beaches in Asia and Central America, reported concentrations reaching 1–730 ng/g for BPA, and 0.1–153 ng/g for t-OCT [6]. BPA has been listed as one of the very high concern compounds by the European Chemicals Agency [7]. The European Food Safety Authority (EFSA) reassessed in 2015 the current tolerable daily intake (TDI) for BPA and reduced it from 50 to a temporary value of 4 μg /kg b. w. daily [8]. However, the plastic industries are in a legal dispute against the European Chemicals Agency's (ECHA) decision to identify bisphenol A (BPA) as a "substance of very high concern" [9]. It appears that, although BPA is banned in some countries such as Canada, the fight for BPA to stay on shelves continues in Europe [10]. On the other hand, t-OCT is under assessment as persistent, bio—accumulative, and toxic, especially to aquatic life [11,12]. Both compounds present endocrine disrupting properties, resulting from their structural resemblance to the human 17β-estradiol [13], with t-OCT presenting higher estrogenicity than the other pollutants [14,15].

Following uptake, BPA is metabolized through its transformation to glucuronide and sulfate derivatives by vertebrates, such as tadpoles and fish [16], as well as mammals [17], while bacteria lead to the biotransformation of BPA to Hydroquinone, 4-Hydroxyacetophenone, 4-Hydroxybenzoic acid, and 4-Isopropenylphenol [18]. Although BPA has been detected in invertebrates, the reports on its metabolism by aquatic invertebrates are available only for bivalves, which transform it to mono- and disulfate [19]. In a similar manner, alkylphenols, including t-OCT, are metabolized by glucuronidation and/or sulfation in the liver of mammals such as rats [20] and humans [21], while bacteria transform it to end products including hydroquinone, 4-Hydroxybenzaldehyde and 4-Hydroxybenzoic acid [22].

Endocrine disruptors, such as BPA, induce oxidative stress and exert toxic effects on freshwater aquatic organisms by altering bacteria composition in their environment [23]. In the marine environment, sorption of EDCs from plastic particles in the surrounding seawater has been documented, indicating that marine plastics, under environmentally induced stress conditions, can act as carriers of organic contaminants and transfer them to marine organisms [3]. Marine organisms ingest the microplastics together with their contaminants. Amounts of BPA have been detected in fish [24], bivalves [25], and seawater [26], while t-OCT was detected in fish [27], mollusks [28], and shrimp [29]. Nowadays, the largest portion of all seafood that are used as food for humans, originate from aquaculture [30]. Aquaculture relies on plastic use in many aspects, such as fish cages, fish feeders, and fish tanks [31,32]. Moreover, aquaculture hatcheries require the live feed *Artemia* nauplii or metanauplii for feeding of the cultured fish and crustacea larvae [33,34], due to its high nutritional value characterized by high contents of neutral lipids [35]. Although *Artemia* is a crustacean adaptable to a wide range of environmental conditions, it is absent in common marine ecosystems, with its natural habitat being high salinity environments such as salt lakes, where it is found in the form of dormant cysts [36]. This saltwater planktonic crustacean, being a filter-feeder, consumes small-sized particles such as microalgae or organic manure [37], and even microplastics that are present in the water column [38,39]. The transfer of pollutants from microplastics to *Artemia* nauplii, and then to the zebrafish that consumed these nauplii as a feed, has been demonstrated for benzopyrene [40]. The genus *Artemia* spp. is widely used as a toxicity testing model due to its handling advantages, which include the fast hatching of cysts to nauplii, metanauplii, or adults, and hence the ease of access to these developmental stages [41,42]. Moreover, BPA toxicity against the first developmental stages (Instar I-II) of *Artemia franciscana* nauplii [43,44] as well as of the alkylphenol n-hexylphenol against *A. sinica* [45] have been reported, without, however, including t-octylphenol or a determination of the compounds in *Artemia*. The transfer of the endocrine disruption across species from the freshwater planktonic crustacean *D. magna* to both its consumer organisms, namely fish [46], as well as to its prey that is the algae it feeds on [47], was considered an indication of a broader action of endocrine disruptors that extends beyond the target organisms, further emphasizing the ecological risk.

Bisphenol-A has gained the focus of attention and the determination of BPA in food [48], and environmental samples [49] has been extensively reviewed. The reported methods mostly employ solvent extraction and SPE (polymer, Oasis, or C18) for the isolation of BPA from samples, followed by HPLC analysis (C18 columns). Moreover, the determination of BPA has been reported in biological samples such as rat tissues by HPLC, employing different extraction protocols for the serum and tissues and using a C18 column and an acetonitrile and water mobile phase with gradient elution [50]. In human breast milk, BPA was extracted using Matrix Solid Phase Dispersion (LiChrolut cartridges) and reversed phase HPLC, followed with isocratic elution of a C18 column with acetonitrile-water (70:30, $v/v$) [51]. The determination of the residual monomers, including BPA, released from resin-based dental restorative materials employed HPLC analysis in a Kromasil 100-C18 column eluted with methanol: acetonitrile: water, 60:15:25%, $v/v$ [52], while the same HPLC analysis in saliva samples was performed using a Perfect Sil Target ODS-3 column eluted with acetonitrile/water, 58/42% $v/v$ [53,54]. Fewer studies are engaged in the determination of multiple endocrine disruptors in complex biological samples, since extensive clean-up due to matrix interferences is required and only trace levels of the compounds are present. To this end, elaborate techniques including microwave-assisted extraction, C18 SPE, derivatization, and GC analysis were employed for the simultaneous determination of steroid EDCs, t-OCT, 4-cumylphenol, 4-nonylphenol, and BPA in fish samples [55]. The determination of BPA either alone or in combination with 4-t-octylphenol and 4-nonylphenol in human media has been thoroughly reviewed, and was performed in blood serum by ELISA or RIA, in plasma and urine by LC equipped with a fluorescence or electrochemical detector and a C18 column, while more complex samples, such as semen and placental tissue, required the use of LC-MS using a Shodex column, or LC-MS/MS using a C18 column [56]. A recent report on the determination of BPA and other xenoestrogens including t-OCT in human urine, serum, and breast milk samples, employed liquid chromatography-tandem mass spectrometry (LC-MS/MS) method on a silica Acquity column [57]. Analysis of several emerging contaminants, including BPA but not t-OCT, was performed in poultry manure using ultrasound-assisted matrix solid-phase dispersion for the extraction of the analytes and analysis following derivatization was performed by GC coupled to tandem mass spectrometry [58]. Simultaneous determination of BPA and alkylphenols has recently been reported for fish and gull [27], water samples [59], mussels [60], and naturally occurring marine zooplankton [61], using ultrasonic extraction followed by SPE (Oasis HLB) and analysis was performed by HPLC-FLD using a HYPERSIL GOLD C18 column and gradient elution with acetonitrile: water. The methods for the simultaneous determination of BPA and certain alkylphenols mentioned above, require specific equipment including ultrasonic extraction, fluorescence, MS or MS/MS detectors, which are not always available. A simple HPLC method, for the determination of BPA and its metabolites in bacterial cultures, was previously developed [18,62] and employed filtering of the medium samples and SPE (C18) extraction of BPA form bacteria, while HPLC analysis was conducted on a C18 Nucleosil column with gradient elution (acetonitrile: water containing TFA). However, this method did not include t-OCT and was validated only regarding the determination of BPA and its metabolites in cultures of bacteria in minimal salt media with BPA as the main carbon source. The live fish feed *Artemia* nauplii and metanauplii are unique regarding their biochemical composition, habitat, and extensive use as a live larval feed in aquaculture. Artemia is not included in the naturally occurring zooplanktonic communities, since it lives in high salinity habitats [36]. Its developmental stages, the nauplii and metanauplii, contain high levels of protein and lipids and especially unsaturated fatty acids and their composition can be enhanced through their enrichment with the appropriate nutrients, a fact that renders them suitable to serve as a live feed of larval stages in aquaculture [63]. However, this particular composition of Artemia as well as of its culture medium, namely seawater 35 ppt, poses significant analysis problems due to matrix interferences. Although the transfer of organophosphorus pesticides [64] and nonylphenol [65] from the crustacean *Artemia* to fish has been documented, the risk of the transfer of BPA and t-OCT to the cultured larvae of marine organisms through their live feed has not been evaluated, possibly due to the scarcity of methods on the determination of bisphenol and t-octylphenol in *Artemia*.

In the present study, *Artemia franciscana* metanauplii were used to develop and validate an analytical method that would enable the quantification of BPA and t-octylphenol, in the organisms as well as in their culture medium. The application of this method demonstrates that both bisphenol and t-octylphenol are ingested by *Artemia* and allows their quantification.

## 2. Materials and Methods

### 2.1. Chemicals and Reagents

The organic solvents, methanol and absolute ethanol, were supplied by Fisher Scientific (Loughborough, UK), while acetonitrile was supplied by VWR Chemicals (Paris, France). Trifluoroacetic acid (TFA) and sodium hypochlorite solution 10% $w/v$ were purchased from AppliChem GmbH (Darmstadt, Germany). The sea salt used as the *Artemia* culture medium was purchased from Instant Ocean (Blacksburg, VA, USA). *Artemia franciscana* cysts were kindly supplied by INVE Aquaculture (INVE HELLAS SA, 93, Kyprou str., 16451 Argyroupoli, Athens, Greece). The $C_{18}$ sorbent material, columns, and frits for SPE were supplied by Grace Davison Discovery Sciences (Bannockburn, IL, USA).

The internal standard 4-n-Octylphenol and Bisphenol-A were supplied by Alfa Aesar (Karlsruhe, Germany), while 4-tert-Octylphenol was obtained from Sigma-Aldrich (Germany). The investigated compounds are presented in Table 1.

**Table 1.** Compounds investigated and characteristic parameters.

| Compound (Abbreviation, CAS#, Formula) | Structure | MW | Solubility in Water (mg L$^{-1}$) | LD50 g kg$^{-1}$ | TDI (Tolerable Daily Intake) ng/kg bw/day |
|---|---|---|---|---|---|
| Bisphenol-A BPA, 80-05-7 $C_{15}H_{16}O_2$ | | 240.20 | 120 [66] | 3-5 (rat) [67] | 4000 (human) [68] |
| 4-tert-Octylphenol t-OCT, 140-66-9 $C_{14}H_{22}O$ | | 206.32 | 5.1 [69] | 4.6 (rat) [69] | 0.067 (men) 33.3 (women) [70] |
| 4-n-Octylphenol n-OCT, 1806-26-4 $C_{14}H_{22}O$ | | 206.32 | 3.1 [71] | 87.8 µg L$^{-1}$ (fish) [71] | - |

### 2.2. Animals

Experiments were performed on *A. franciscana* (Kellogg) cysts (e.g., grade, GSL strain) provided by INVE (INVE HELLAS S.A., Athens, Greece). The cysts were stored at 4 °C until use. The axenic culture of *Artemia* cysts was performed following their hydration and decapsulation in hypochlorite solution, as previously described [72,73]. In all procedures the artificial sweater was autoclaved prior to use and the containers were rinsed with ethanol to ensure bacteria-free cultures. The nauplii (instar I) were cultured for a total of 96h at a density of 15 nauplii per mL and they were fed on processed yeast provided by P. Sorgeloos (Laboratory of Aquaculture and *Artemia* Reference Center, University of Ghent, Belgium). The sterile culture medium was renewed daily, and feeding was stopped prior to the addition of pollutants at 72 h. In the preliminary experiments, the LC$_{50}$ was determined for each endocrine disruptor. To this end, two experimental series of nauplii were employed at 72 h since the onset of cyst incubation after being fasted for six hours, and the first series received increasing concentrations of bisphenol (0, 5, 10, 15, 20, 25 or 30 µg/mL) while the second series received increasing concentration of 4-tert-octylphenol (0, 0.25, 0.5, 0.75, 2, 4 or 6 µg/mL). Each series employed samples containing 0 µg/mL of endocrine disruptor to serve as controls. Survival and mortality were recorded following a 24 h exposure to the pollutant and $LC_{50}$ was estimated using linear regression and Probit analysis.

### 2.3. Preparation of Stock and Standard Solutions

The stock solution of BPA (500 µg/mL) and of t-OCT (200 µg/mL) and of the internal standard n-octylphenol (500 µg/mL) were prepared in methanol. All stock solutions were kept in 4 °C, in glass volumetric flasks protected from light to avoid photodegradation of the analytes.

A set of six working standard solutions were prepared for BPA in two sets containing 2.5, 5, 7.5, 10, 12.5 and 15 µg/mL or 1, 25, 50, 75, 100 and 125 µg/mL. For t-OCT a set of six working standard solutions were prepared containing 1, 2.5, 5, 7.5 10 and 12.5 µg/mL. Each spiked sample contained the internal standard n-octylphenol at a concentration of 75 µg/mL. The standard solutions were transferred in amber glass vials (SU860083, Supelco) and stored at 4 °C. Calibration curves were prepared at six points, with concentrations ranging from 2.5 to 12.5 µg/mL for BPA in tissue, from 1 to 125 µg/mL for BPA in medium and from 1 to 12.5 µg/mL for t-OCT.

### 2.4. Sample Preparation

The protocol previously developed for the extraction of Bisphenol-A in bacterial cultures [61] was modified appropriately and validated to facilitate purification of both BPA and t-OCT in *Artemia* nauplii, as well as in their culture medium.

Regrading extraction of BPA and t-OCT from biological samples, 0.2 g wet weight of *A. franciscana* nauplii were homogenized (4 °C) in 2 mL methanol and 10.5 mL of double distilled water containing 0.08% (v/v) TFA, following the addition of the internal standard n-octylphenol (final concentration of 75 µg/mL). A Kinematica Polytron PCU homogenizer was employed for 5 min and the sample was kept in ice. The homogenate was centrifuged for 10 min at 10,000 rpm. The supernatant was submitted to SPE (0.5 g of $C_{18}$ sorbent) conditioned with 10 mL methanol, 5 mL ddH$_2$O, and 5 mL ddH$_2$O containing 0.08% TFA. Loading of the samples (manual flow 1 mL/min) followed and columns were washed with 5 mL ddH$_2$O. Elution of the compounds was performed with 4 mL methanol and the eluates were evaporated to dryness (rotary evaporator, 40 °C). Acetonitrile (1 mL) was added to the dried samples, and they were stored at −20 °C, protected from light. During the analyses all tubes were protected from light to avoid photodegradation of the analytes BPA and 4-tOP.

For the extraction of the analytes from the culture medium of *Artemia* nauplii, a 10 mL sample of the culture medium was used, its pH value was adjusted to 3.0 ± 0.1 using a 10% water solution of TFA (v/v) and the internal standard was added (75 µg/mL). Following the addition of 2 mL methanol, the sample was centrifuged for 10 min at 10,000 rpm and the supernatant was submitted to SPE, as described for biological samples.

### 2.5. Pollutant Administration

For the feeding assays, nauplii cultures (1000 mL cultures in filtered and autoclaved artificial seawater, salinity 35 ppt) [72,73] were fasted for 6 h and then Bisphenol-A or 4-tert-octylphenol was administered at 72 h, at a final concentration of each substance in each culture of 10 µg/mL and 0.5 µg/mL, respectively. The cultures were then incubated for 24 h under natural light. At the end of the incubation the nauplii and their culture medium were separated by filtration and stored at −20 °C.

*2.6. Photodegradation Assay*

Experiments on the potential photodegradation of the analytes were performed in glass tubes containing sterile seawater and BPA or t-OCT at a concentration of 50 µg/mL and 10 µg/mL, respectively. The tubes were incubated under intense sunlight for a total of 48 h. The concentration of the analytes BPA and t-OCT was determined at the onset (t = 0), at t = 24 h and at the end of the incubation period (t = 48 h), following extraction of the analytes.

*2.7. Chromatography*

The HPLC system comprised of an LC20$_{AD}$ pump and an SPD-20A photodiode array detector (DAD) (Shimadzu, Kyoto, Japan), a Rheodyne 7125 injection valve, with a loop of 80 µL volume (Rheodyne, Cotati, CA, USA) and a Nucleosil 100 C$_{18}$ column (250 × 4.6 mm, 5 µm), (Macherey-Nagel GmbH & Co., Duren, Germany) equipped with a 10 × 4.6 mm I.D., Nucleosil C$_{18}$ precolumn. Separation of BPA, t-OCT, and n-OCT was performed using a binary mobile phase system of mobile Phase A (95:5, $v/v$, 0.1 % TFA in H$_2$O: Acetonitrile) and mobile Phase B (5:95, $v/v$, 0.08 % TFA in H$_2$O: Acetonitrile) at room temperature, as previously reported [62] but with adjustments to the gradient, starting at 100% mobile Phase A and proceeding to an increasing mobile Phase B concentration to 20% over 2 min, 70% over 5 min, 90% at 7 min, and until the end of the analysis. Flow rate was 0.5 mL/min, and the analytes were detected at 220 nm.

## 3. Results and Discussion

*3.1. Analytes Extraction Efficiency*

*Artemia* metanauplii are a solid sample that require homogenization to achieve cell lysis, and this was performed by homogenization of the sample in methanol-H$_2$O containing 0.08% ($v/v$) TFA and centrifugation. The previously described protocol for the extraction of BPA from bacterial cultures [62] was modified regarding the methanol aqueous phase ratio from 5:1 to about 1:3, since this amount of methanol produced better extraction of analytes. Since alkylphenols are characterized by low solubility in water compared to Bisphenol-A (Table 1), methanol contributed to the increased solubility of t-octylphenol and BPA. The use of ethyl acetate or water, instead of methanol-TFA, was also employed in trial experiments but compound recovery was deemed inadequate as it resulted in the coextraction of *Artemia* lipids and matrix interferences. In the present study, 0.2 g wet weight tissue, corresponding to 0.01 ± 0.003 g dry weight, were processed since they resulted in satisfactory chromatograms with no interfering peaks in the blank samples (Figure S1). Although microwave-assisted solvent extraction [74], as well as ultrasonic bath extraction [75], were reported for the extraction of bisphenol and alkylphenols, homogenization is common in biological samples [75,76].

Following the first methanol-water-TFA extraction step, the samples were subjected to SPE, an acknowledged technique for the selective sample preparation prior to analysis of endocrine disruptors in liquid or pre-extracted solid samples [75]. SPE has been previously used for the extraction of bisphenol from complex matrices, such as rat tissues [50], human serum [77], or bacterial cultures [18]. The commonly used sorbent is C18 [18,50,77] although the use of polymeric sorbents [59] and silica gel [78] have been reported. However, C18 cartridges provided the cleanest extracts of fish samples compared to polymeric sorbents, employed in the analysis of bisphenol and alkylphenols [55]. A flow rate of 1 mL/min during the elution of SPE, and a water bath temperature lower than 40 °C during the evaporation of methanol, after SPE, were employed as previously recommended for bacteria samples [62] to facilitate the effective recovery of both analytes and the internal standard. However, it should be noted that although simple centrifuging and filtering was sufficient for medium samples originating from bacterial cultures in basal salt medium [62], this was not the case with the *Artemia* medium samples, since it resulted in clogging of the injection loop, due to the higher salt concentration of the sample. Hence, the SPE step was critical for efficient analysis.

The extraction in water-methanol and the following SPE step resulted in excellent recovery of BPA (100.4 ± 3.1) and of t-OCT (100.6 ± 3.9) in the case of the *Artemia* samples. The mean recovery in culture medium samples amounted to 99.9 ± 0.8 for BPA, while for t-OCT it amounted to 101.7 ± 1.9 (Table 2). The recovery values reported in the present work are higher to the previously reported values of 83.7 % (BPA) and 87.4 % (t-OP), for water extraction of Bisphenol-A and alkylphenols from zooplankton samples using Oasis HLB glass cartridges [27]. Our results for BPA and t-OCT extraction are in accordance with a recently published report on a water-acetonitrile extraction of bisphenols and octylphenols from fish and solid matrices, in which affinity chromatography clean-up instead of the reversed phase $C_{18}$-SPE was employed and lower concentrations of 1 ng/g were used [79], resulting in recoveries of up to 101.1% for BPA and to 87.9 for t-octylphenol.

Table 2. Linearity, accuracy (bias), precision, and recovery of BPA and t-OCT determination in the tissue and the culture medium of *Artemia*.

| Analyte Sample Std. Curve /$R^2$ | Nominal Conc. (µg/mL) | Calculated Conc. (µg/mL) (Mean ± SD) | Relative Bias (%) | Precision | | Recovery (%) | |
|---|---|---|---|---|---|---|---|
| | | | | Intra- ($n = 3$) | Inter- ($n = 2 \times 3$) | (Mean ± SD) | RSD% |
| BPA culture medium $y = 0.0027x - 0.024$ 0.9998 | 1 | 0.94 ± 0.002 | −5.3 | 0.2 | 0.3 | 94.8 ± 0.3 | 0.3 |
| | 25 | 24.86 ± 0.1 | −0.6 | 0.4 | 0.5 | 99.4 ± 0.4 | 0.4 |
| | 50 | 50.6 ± 0.3 | 1.2 | 0.6 | 0.7 | 101.2 ± 0.6 | 0.6 |
| | 75 | 74.8 ± 0.09 | −0.2 | 0.1 | 0.2 | 99.8 ± 0.1 | 0.1 |
| | 100 | 100.2 ± 0.6 | 0.2 | 0.6 | 0.7 | 100.2 ± 0.6 | 0.6 |
| | 125 | 123.7 ± 1.8 | −1.1 | 0.4 | 1.4 | 98.9 ± 1.4 | 1.4 |
| BPA Artemia $y = 0.045x - 0.0887$ 0.999 | 2.5 | 2.64 ± 0.02 | 5.9 | 0.4 | 0.6 | 105.6 ± 0.6 | 0.6 |
| | 5 | 4.81 ± 0.05 | −3.8 | 0.8 | 1.0 | 96.3 ± 0.9 | 0.9 |
| | 7.5 | 7.61 ± 0.02 | 1.3 | 0.1 | 0.3 | 101.4 ± 0.3 | 0.3 |
| | 10 | 9.86 ± 0.02 | 0.1 | 0.2 | 0.7 | 98.6 ± 0.2 | 0.2 |
| | 12.5 | 12.45 ± 0.03 | −0.4 | 0.1 | 0.2 | 99.6 ± 0.2 | 0.2 |
| | 15 | 15.11 ± 0.02 | 0.7 | 0.03 | 0.1 | 100.7 ± 0.1 | 0.1 |
| t-OCT culture medium $y = 0.018x + 0.037$ 0.999 | 1 | 1.03 ± 0.04 | 3.1 | 0.6 | 4.3 | 103.1 ± 4.5 | 4.3 |
| | 2.5 | 2.60 ± 0.04 | 4.1 | 0.5 | 1.6 | 104.1 ± 1.7 | 1.6 |
| | 5 | 5.05 ± 0.03 | 1.1 | 0.03 | 0.7 | 101.1 ± 0.7 | 0.7 |
| | 7.5 | 7.69 ± 0.02 | 2.5 | 0.1 | 0.2 | 102.5 ± 0.2 | 0.2 |
| | 10 | 9.87 ± 0.03 | −1.2 | 0.1 | 0.3 | 98.8 ± 0.3 | 0.3 |
| | 12.5 | 12.75 ± 0.1 | 2.0 | 0.7 | 0.9 | 102.0 ± 0.9 | 0.9 |
| t-OCT Artemia $y = 0.019x + 0.012$ 0.9989 | 1 | 1.08 ± 0.02 | 8.7 | 0.01 | 1.9 | 108.7 ± 2.0 | 1.9 |
| | 2.5 | 2.41 ± 0.01 | −3.4 | 0.5 | 0.6 | 96.6 ± 0.6 | 0.6 |
| | 5 | 4.85 ± 0.09 | −2.8 | 0.4 | 1.8 | 97.2 ± 1.8 | 1.8 |
| | 7.5 | 7.59 ± 0.03 | 1.3 | 0.1 | 0.4 | 101.3 ± 0.4 | 0.4 |
| | 10 | 9.98 ± 0.02 | −0.2 | 0.3 | 0.3 | 99.8 ± 0.2 | 0.2 |
| | 12.5 | 12.49 ± 0.2 | 0.2 | 0.4 | 1.8 | 100.2 ± 1.8 | 1.8 |

*3.2. Method Validation*

The results of the linear regression analysis performed are presented below (Table 2). The use of an internal standard was deemed necessary since sample matrix complexity, due to the presence of proteins and lipids in both samples, required an extraction protocol for both tissues and culture medium.

To this end, 4-n-octylphenol was selected, since it has a similar structure to both bisphenol and tert-octylphenol, is well resolved from both analytes, and is not present in the blank samples (Figure S1).

Linearity of the standard curves in the *Artemia* culture medium and *Artemia* metanauplii (Table 2) was excellent, as indicated by the $R^2$ values of 0.999 for BPA and t-OCT. The Relative Bias estimated as the percentage of ((calculated- nominal)/nominal]) was within the accepted range of ±10%. Intra-day precision was calculated by three analyses performed on the same day, while inter-day precision was calculated by performing six analyses over three days and estimated as % RSD, never exceeding 5.0% for all analytes. Our results indicate that the method is precise and repeatable with excellent recovery rates. The nominal concentrations of 1.0 µg/mL BPA and 1.0 µg/mL t-OCT in the culture medium were below the calculated LOD values (Table 3) and were, thus, excluded from the estimation of recovery results described above. The values for LOD and LOQ calculation were based on the response and the slope of the calibration curve.

Table 3. LOD, LOQ, and RT values for BPA and t-OCT.

| Std. Curve | Retention Time (min) Mean ± SD (RSD %) | LOD (µg/mL) | LOQ (µg/mL) |
|---|---|---|---|
| BPA in *Artemia* | 15.200 ± 0.12 (0.82) | 0.21 | 0.65 |
| BPA in culture medium | 15.310 ± 0.10 (0.69) | 0.63 | 1.92 |
| t-OCT in *Artemia* | 21.876 ± 0.11 (0.49) | 0.17 | 0.52 |
| t-OCT in culture medium | 21.597 ± 0.31 (1.46) | 0.41 | 1.25 |

The value of a LOD of 0.21 µg/mL and 0.17 µg/mL were estimated for BPA and t-OCT, respectively, in *Artemia* tissue corresponding to 10.5 ng/g dry weight and 8.6 ng/g dry weight, respectively, and of 0.63 µg/mL and 0.41 µg/mL for *Artemia* culture medium (Table 3). Although LOD values of 0.07 µg/kg were reported for BPA in mussels' tissue [80], estimations were made with σ taken as the standard deviation of three samples spiked at the experimental estimated LOQ, and S as the slope of the corresponding calibration curve. On the other hand, LOD values of 5.6 µg/mL and 6.3 µg/mL, were reported following liquid–liquid extraction of water samples [81]. The previously reported LOQ values in zooplankton amounted to 2 ng/g dry weight for BPA and 0.8 ng/g dry weight for t-OCT, and to 5 and 1 ng/L, respectively, for water samples, determined as a tenfold signal-to-noise ratio for a sample with a very low analyte content [27], employing different methods for the biological and water samples. Variable results were obtained from the computation of LOD and LOQ values, depending on the statistical approach used. The residual standard deviation or error of intercept of calibration line was recommended instead of the mean blank signal value, as it is considered that it provides a more accurate estimate [82].

Since the LOD and LOQ values are calculated based on the calibration curve, they greatly depend on its range and lower values are expected when a range of lower concentrations are used. However, the range of the calibration curve should extend over the range of expected analytes' concentrations in the samples [83]. In the present work, the expected concentrations of Bisphenol-A and t-octylphenol were within the range of the $LC_{50}$ values. These values were acquired from the toxicity experiments that allowed the estimation of the amount that could be administered to *Artemia* metanauplii without leading to high mortality rates.

Typical HPLC chromatograms of *Artemia* and culture medium samples spiked with BPA, t-OCT, and n-OCT are presented in Figure 1. The peaks of the analytes BPA, t-OCT, and of the internal standard n-OCT, showed with a retention time (Rt) of 15.200 ± 0.12, 21.876 ± 0.11 26.241 ± 0.1 min, respectively. The peak appearing from 0 to 6 min, in all samples, possibly corresponds to the solvents, while the peak appearing at approximately 7.0 ± 0.5 min only in the medium samples possibly corresponds to medium constituents. Similar peaks also appear in the blank samples without, however, interfering with analysis (Figure S1).

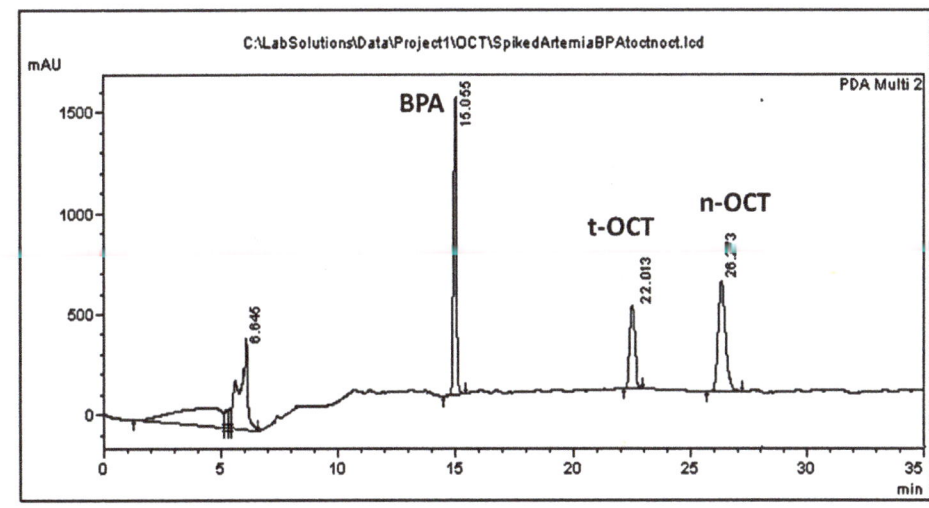

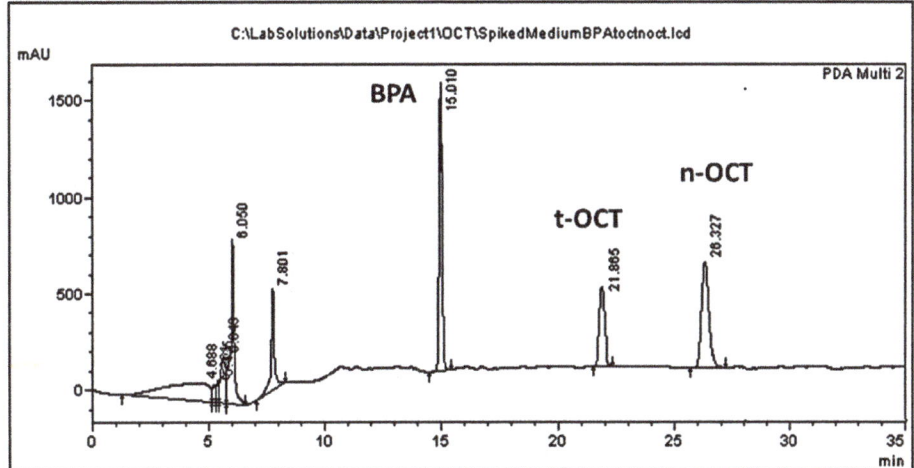

**Figure 1.** HPLC chromatograms of (**A**) *Artemia* tissue; and (**B**) culture medium samples containing BPA (50 μg/mL), t-OCT (10 μg/mL) and the internal standard n-OCT (75 μg/mL).

*3.3. Administration of Pollutants*

BPA was administered at a concentration of 10 μg/mL, while t-OCT was administered at 0.5 μg/mL, since their LOD values amounted to 15.6 and 0.73 μg/mL, respectively (Figure S2). The simultaneous administration of the two pollutants at $LC_{50}$ levels was not applicable, since it resulted in higher mortality values, possible due to synergistic effects. Hence, each pollutant, namely BPA or t-OCT, was administered separately. Acute toxicity and inhibition of growth of the early developmental stages of *Artemia* nauplii (Instar I-II) have been reported after a 24 h exposure to BPA with an $LC_{50}$ of 44.8 mg/L [43] for *Artemia franciscana*. The $LC_{50}$ values for *Artemia sinica* nauplii amounted to 47.5 mg/L for BPA and to 2.27 mg/L for t-OCT [84]. On the other hand, the $LC_{50}$ values of 70.1 μg/L and 7.7 μg/L were reported for adult (15 days old) *Artemia sinica* individuals exposed to BPA and n-hexylphenol [45], illustrating that the advanced developmental stages present greater sensitivity to these pollutants. The increased sensitivity of subsequent developmental stages to pollutants was attributed to the fact that the organisms appear more ravenous compared to the earlier developmental stages, during which the animals depend on reserved yolk for their energy requirements [85]. In addition to the importance of the developmental stage employed, the origin of cysts and abiotic parameters regarding, among others, the pH, hatching temperature, photoperiod duration, and saltwater treatments are also of great importance [42]. Considering that bacteria are capable of BPA biotransformation [18], the use of axenic cultures in the autoclaved medium were used in the present study. Representative chromatograms of the pollutants following administration of 10 µg/mL BPA or 0.5 µg/mL t-OCT to *Artemia*, that are below their estimated $LC_{50}$ values, are presented in Figure 2 for *Artemia*, and in Figure 3 for its culture medium.

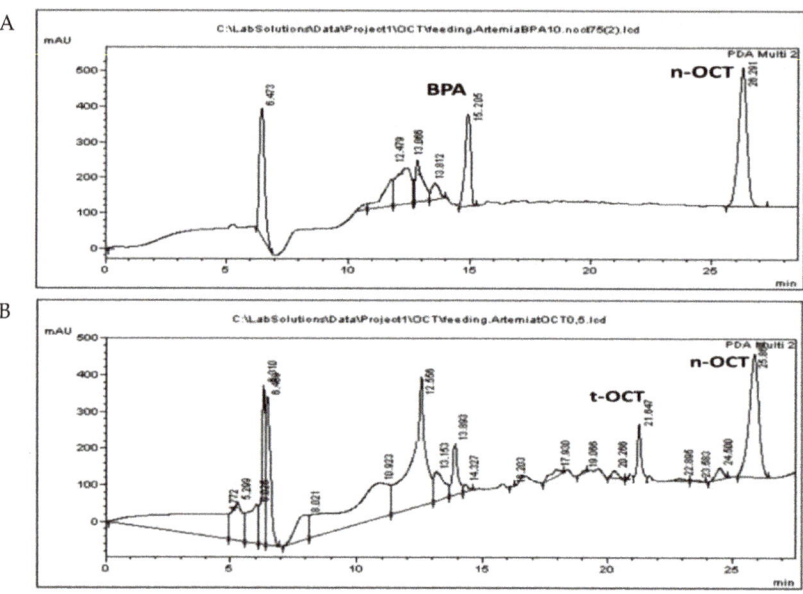

**Figure 2.** HPLC chromatograms of *Artemia* tissue following administration of (**A**) 10 µg/mL BPA; and (**B**) 0.5 µg/mL t-OCT.

The peaks appearing at 12.5 ± 0.18, 13.1 ± 0.08 and 13.8 ± 0.2 and apparent following administration of either pollutant (Figure 2A,B) might correspond to products produced during their biotransformation by *Artemia*. Similar peaks were observed following BPA microbial biotransformation corresponding to 4-Hydroxyacetophenone, 4-Hydroxybenzoic acid, and hydroquinone [18]. Moreover, biodegradation of tert-octylphenol by the fungus *Thielavia* sp. HJ22 has been reported to result in the formation of 4-hydroxybenzoic acid hydroquinone [86]. Although UV spectra presented a good match with standard compounds, the identity of these peaks was not confirmed by MS in the present work. The peak that appears at 19.7 (Figures 2B and 3B) is evident only following administration of t-OCT and might correspond to a photodegradation product, since similar peaks were observed during the photodegradation experiment (Figure S3D). Photodegradation via photolysis has been confirmed for BPA [25,27] and for t-OCT [87], and the products of solar transformation presented higher toxicity than the parent compound [88]. The feeding experiments were conducted under normal light and, hence, photodegradation losses were estimated to examine whether any pollutant concentration decline is due to photolysis. In the photolysis experiments, the initial BPA and t-OCT concentration were calculated at 49.15 µg/mL and 9.81 µg/mL, respectively, and their concentration following a 24 h incubation amounted to 44.63 µg/mL and 7.87 µg/mL, while after 48 h BPA amounted to 37.94 and t-OCT to only 1.7 µg/mL. Hence, the reduction in the analyte concentration induced by photodegradation was 9.2% for BPA and 19.7% for t-OCT at 24 h (Table 4).

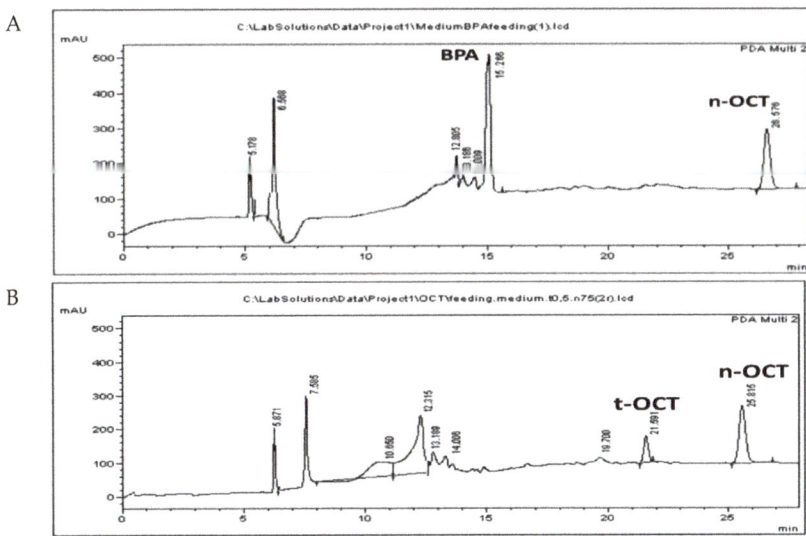

**Figure 3.** HPLC chromatograms of culture medium following the administration of (**A**) 10 µg/mL BPA; and (**B**) 0.5 µg/mL t-OCT.

**Table 4.** Bisphenol A and t-octylphenol levels in administration experiments.

| Compound | BPA | t-OCT |
| --- | --- | --- |
| Administered in 1000 mL culture (µg) | 10,000.0 | 500.0 |
| Found in *Artemia* (µg/g wet weight) | 68.3 | 5.9 |
| Found in culture medium (µg/mL) | 6.68 | 0.3 |
| Total found in *Artemia* (µg) | 40.9 | 3.6 |
| Total found in 1000 mL medium (µg) | 6680 | 319.3 |
| Total found (µg) | 6720.9 | 322.9 |
| Concentration reduction in cultures (%) | 32.8 | 35.4 |
| Decline due to Photolysis (%) | 9.2 | 19.7 |

These values are in accordance with those reported for bisphenol A degradation [89] but higher than the previously reported absence of t-OCT degradation under one hour of direct solar irradiation [90]. Although part of the pollutants is degraded via photolysis, this does not compensate for the observed reduction in the total levels of both pollutants (Table 4). The amounts detected in the processed medium samples were 66.8 and 3.0 µg and the total amounts in a 1000 mL culture corresponded for BPA to 66.8% and for t-OCT to 63.8% of the initially added amounts. In addition, the amounts detected in *Artemia* tissue amounted to 4.1% and 0.7% of the initially added amounts for BPA and t-OCT, respectively, and a reduction of up to approximately 32% in total BPA and of 35% in total t-OCT was observed in our experiments. The accumulation of BPA in zooplankton has been confirmed in an eco-system bioreactor during incubation of phytoplankton (*Nannochloropsis* sp.) and zooplankton (*Artemia* sp. or *Brachinous* sp.) in the presence of the pollutant [91].

The occurrence of high levels of BPA and alkylphenols in the marine environment has been documented in aquaculture facilities which are in the vicinity of industrial or urban activity [92]. The BPA values amounted to 37 ng/L in mussels cultured in Thailand [92], to 1.5 ng/g d. w. in fish cultured in Taiwan [93], to 4.2 ng/g in the muscle of fish cultured in Malaysia [94], or 272 ng/g d. w. in the North Atlantic Ocean [24]. The relevant t- octylphenol concentration was 16 ng/ g d. w. in mussels cultured in Malaysia [94], while the occurrence of the alkylphenol nonylphenol has been documented in the sediments of the Northern Aegean, indicating a risk for the organisms that live in it [95]. The cultured fish fry fed on *Artemia* have a weight of 15 mg and consume about 300 nauplii per day [96], corresponding to a wet weight of 5 mg, indicating that these pollutants, if present in *Artemia*, may accumulate and possibly pose a health risk to the fish larvae. Recently the accumulation of bisphenols in farmed seabream muscle following exposure to microplastics, amounting to 0.3 µg/g muscle, was reported and the authors consider aquaculture equipment as a source of such microplastics [97]. Considering the TDI for BPA amounting to 0.04 ng/kg b. w. daily and the suggested TDI for t-OCT for men of 0.067 ng/kg b. w. daily (Table 1), it becomes apparent that the exposure of cultured organisms to the pollutants might also pose a possible hazard to human health. The use of marine phytoplanktons and zooplanktons has been suggested for the recovery of BPA from seawater [91]. The uncompensated decline in both BPA and t-OCT levels observed in the present study indicates that, although these compounds might accumulate in *Artemia*, their biotransformation is possible. Since the toxicity of the possible biotransformation products has not been investigated and *Artemia* is used as a live feed for cultured marine larvae, the monitoring of its pollutants levels is of extreme importance.

## 4. Conclusions

The developed analytical method provides a simple approach for the simultaneous detection and quantification of bisphenol A and t-octylphenol in the live fish feed *Artemia* and in its culture medium. The presented extraction protocol and the use of 4-n-octylphenol as the internal standard provided a detection limit of 0.21 for BPA and 0.17 ng/µL for t-OCT. The method was applied in the analysis following administration of the pollutants to *Artemia* metanauplii at levels close to their corresponding $LC_{50}$. The administration experiments revealed that *Artemia* is capable to uptake bisphenol and t-octylphenol from the culture medium. Experimental efforts to reduce the pollutant load of *Artemia* are currently under investigation. Further experiments are needed to evaluate the transfer of the pollutants to the fish larvae that feed on *Artemia*.

**Supplementary Materials:** The following supporting information can be downloaded at: https://www.mdpi.com/article/10.3390/mps5030038/s1, Figure S1. HPLC chromatograms of blank *Artemia* tissue (A) and culture medium (B) samples; Figure S2. Linear Regression analysis for the determination of $LC_{50}$ values of BPA and t-OCT; Figure S3. Photodegradation of BPA and t-OCT. HPLC chromatograms of medium with (A)BPA at 0 h; (B) BPA at 48 h; (C) t-OCT at 0 h; (D) t-OCT at 48 h.

**Author Contributions:** M.T. and V.F.S. conceived and designed the experiments; M.T., writing—original draft preparation; V.F.S., review and editing; D.G., formal analysis in BPA experiments; K.D., formal analysis in the t-OCT experiments. All authors have read and agreed to the published version of the manuscript.

**Funding:** This research received no external funding.

**Institutional Review Board Statement:** Not applicable.

**Informed Consent Statement:** Not applicable.

**Acknowledgments:** No additional funding was received in support of the present research work. The authors wish to thank INVE HELLAS S.A., for the provision of *Artemia* cysts and Sorgeloos for the provision of yeast.

**Conflicts of Interest:** The authors declare no conflict of interest.

## References

1. Koelmans, A.A.; Besseling, E.; Foekema, E.M. Leaching of plastic additives to marine organisms. *Environ. Pollut.* **2014**, *187*, 49–54. [CrossRef] [PubMed]
2. Darbre, P.D. Chemical Components of Plastics as Endocrine Disruptors: Overview and Commentary. *Birth Defects Res.* **2020**, *112*, 1300–1307. [CrossRef] [PubMed]
3. Chen, Q.; Allgeier, A.; Yin, D.; Hollert, H. Leaching of endocrine disrupting chemicals from marine microplastics and mesoplastics under common life stress conditions. *Environ. Int.* **2019**, *130*, 104938. [CrossRef] [PubMed]
4. López-Cervantes, J.; Sánchez-Machado, D.; Paseiro-Losada, P.; Simal-Lozano, J. Effects of Compression, Stacking, Vacuum Packing and Temperature on the Migration of Bisphenol A from Polyvinyl Chloride Packaging Sheeting into Food Simulants. *Chromatographia* **2003**, *58*, 327–330. [CrossRef]
5. Hahladakis, J.N.; Velis, C.A.; Weber, R.; Iacovidou, E.; Purnell, P. An overview of chemical additives present in plastics: Migration, release, fate and environmental impact during their use, disposal and recycling. *J. Hazard. Mater.* **2018**, *344*, 179–199. [CrossRef]
6. Hirai, H.; Takada, H.; Ogata, Y.; Yamashita, R.; Mizukawa, K.; Saha, M.; Kwan, C.; Moore, C.; Gray, H.; Laursen, D.; et al. Organic micropollutants in marine plastics debris from the open ocean and remote and urban beaches. *Mar. Pollut. Bull.* **2011**, *62*, 1683–1692. [CrossRef]
7. European Chemicals Agency (ECHA). Candidate List of Substances of Very High Concern for Authorization (2020). Available online: https://echa.europa.eu/candidate-list-table (accessed on 3 April 2022).
8. EFSA. Bisphenol, A. 2017. Available online: https://www.efsa.europa.eu/en/topics/topic/bisphenol (accessed on 2 April 2022).
9. Available online: https://plasticseurope.org/wp-content/uploads/2021/10/20210301-BPA-Judgment-appeal.pdf (accessed on 1 April 2022).
10. Vogel, S.A. The politics of plastics: The making and unmaking of bisphenol a "safety". *Am. J. Public Health* **2009**, *99* (Suppl. S3), S559–S566. [CrossRef]
11. Ying, G.G.; Williams, B.; Kookana, R. Environmental fate of alkylphenols and alkylphenol ethoxylates—A review. *Environ. Int.* **2002**, *28*, 215–226. [CrossRef]
12. Octylphenol. Available online: https://echa.europa.eu/el/substance-information/-/substanceinfo/100.060.634 (accessed on 2 April 2022).
13. Kang, J.H.; Kondo, F.; Katayama, Y. Human exposure to bisphenol A. *Toxicology* **2006**, *226*, 79–89. [CrossRef]
14. White, R.; Jobling, S.; Hoare, S.; Sumpter, J.; Parker, M. Environmentally persistent alkylphenolic compounds are estrogenic. *Endocrinology* **1994**, *135*, 175–182. [CrossRef]
15. Olaniyan, L.; Okoh, O.O.; Mkwetshana, N.T.; Okoh, A.I. Environmental Water Pollution, Endocrine Interference and Ecotoxicity of 4-tert-Octylphenol: A Review. *Rev. Environ. Contam. Toxicol.* **2020**, *248*, 81–109. [CrossRef]
16. Canesi, L.; Fabbri, E. Environmental Effects of BPA: Focus on Aquatic Species. *Dose Response* **2015**, *13*, 1559325815598304. [CrossRef] [PubMed]
17. Nachman, R.M.; Hartle, J.C.; Lees, P.S.; Groopman, J.D. Early Life Metabolism of Bisphenol A: A Systematic Review of the Literature. *Curr. Environ. Health Rep.* **2014**, *1*, 90–100. [CrossRef] [PubMed]
18. Kyrila, G.; Katsoulas, A.; Schoretsaniti, V.; Rigopoulos, A.; Rizou, E.; Doulgeridou, S.; Sarli, V.; Samanidou, V.; Touraki, M. Bisphenol A removal and degradation pathways in microorganisms with probiotic properties. *J. Hazard. Mater.* **2021**, *413*, 125363. [CrossRef]
19. Hayashi, O.; Kameshiro, M.; Masuda, M.; Satoh, K. Bioaccumulation and metabolism of [14C] bisphenol A in the brackish water bivalve *Corbicula japonica*. *Biosci. Biotechnol. Biochem.* **2008**, *72*, 3219–3224. [CrossRef]
20. Certa, H.; Fedtke, N.; Wiegand, H.J.; Müller, A.M.; Bolt, H.M. Toxicokinetics of p-tert-octylphenol in male Wistar rats. *Arch. Toxicol.* **1996**, *71*, 112–122. [CrossRef]
21. Jing, X.; Bing, S.; Xiaoyan, W.; Xiaojie, S.; Yongning, W. A study on bisphenol A, nonylphenol, and octylphenol in human urine samples detected by SPE-UPLC-MS. *Biomed. Environ. Sci.* **2011**, *24*, 40–46. [CrossRef]
22. Rajendran, R.K.; Huang, S.L.; Lin, C.C.; Kirschner, R. Biodegradation of the endocrine disrupter 4-tert-octylphenol by the yeast strain *Candida rugopelliculosa* RRKY5 via phenolic ring hydroxylation and alkyl chain oxidation pathways. *Bioresour. Technol.* **2017**, *226*, 55–64. [CrossRef]
23. Pop, C.-E.; Draga, S.; Măciucă, R.; Niță, R.; Crăciun, N.; Wolff, R. Bisphenol A Effects in Aqueous Environment on *Lemna minor*. *Processes* **2021**, *9*, 1512. [CrossRef]
24. Barboza, L.; Cunha, S.C.; Monteiro, C.; Fernandes, J.O.; Guilhermino, L. Bisphenol A and its analogs in muscle and liver of fish from the North East Atlantic Ocean in relation to microplastic contamination. Exposure and risk to human consumers. *J. Hazard. Mater.* **2020**, *393*, 122419. [CrossRef]
25. Baralla, E.; Pasciu, V.; Varoni, M.V.; Nieddu, M.; Demuro, R.; Demontis, M.P. Bisphenols' occurrence in bivalves as sentinel of environmental contamination. *Sci. Total Environ.* **2021**, *785*, 147263. [CrossRef] [PubMed]
26. Staples, C.; van der Hoeven, N.; Clark, K.; Mihaich, E.; Woelz, J.; Hentges, S. Distributions of concentrations of bisphenol A in North American and European surface waters and sediments determined from 19 years of monitoring data. *Chemosphere* **2018**, *201*, 448–458. [CrossRef] [PubMed]

27. Staniszewska, M.; Falkowska, L.; Grabowski, P.; Kwaśniak, J.; Mudrak-Cegiołka, S.; Reindl, A.R.; Sokołowski, A.; Szumiło, E.; Zgrundo, A. Bisphenol A, 4-tert-octylphenol, and 4-nonylphenol in the Gulf of Gdańsk (Southern Baltic). *Arch. Environ. Contam. Toxicol.* **2014**, *67*, 335–347. [CrossRef] [PubMed]
28. Salgueiro-Gonzalez, N.; Turnes-Carou, I.; Besada, V.; Muniategui-Lorenzo, S.; Lopez-Mahia, P.; Prada-Rodriguez, D. Occurrence, distribution and bioaccumulation of endocrine disrupting compounds in water, sediment and biota samples from a European river basin. *Sci. Total Environ.* **2015**, *529*, 121–130. [CrossRef] [PubMed]
29. Diao, P.; Chen, Q.; Wang, R.; Sun, D.; Cai, Z.; Wu, H.; Duan, S. Phenolic endocrine-disrupting compounds in the Pearl River Estuary: Occurrence, bioaccumulation and risk assessment. *Sci. Total Environ.* **2017**, *584–585*, 1100–1107. [CrossRef] [PubMed]
30. Little, D.C.; Newton, R.W.; Beveridge, M.C. Aquaculture: A rapidly growing and significant source of sustainable food? Status, transitions and potential. *Proc. Nutr. Soc.* **2016**, *75*, 274–286. [CrossRef] [PubMed]
31. Lusher, A.L.; Hollman, P.C.H.; Mendoza-Hill, J.J. *Microplastics in Fisheries and Aquaculture: Status of Knowledge on Their Occurrence and Implications for Aquatic Organisms and Food Safety*; FAO Fisheries and Aquaculture Technical Paper. No. 615; Food and Agriculture Organization of the United Nations: Rome, Italy, 2017; ISBN 978-92-5-109882-0. Available online: https://www.fao.org/3/i7677e/i7677e.pdf (accessed on 20 March 2022).
32. Vázquez-Rowe, I.; Ita-Nagy, D.; Kahhat, R. Microplastics in fisheries and aquaculture: Implications to food sustainability and safety. *Curr. Opin. Green Sust.* **2021**, *29*, 100464. [CrossRef]
33. Kolkovski, S.; Curnow, J.; King, J. Intensive rearing system for fish larvae research II. *Artemia* hatching and enriching system. *Aquac. Eng.* **2004**, *31*, 309–317. [CrossRef]
34. Turcihan, G.; Turgay, E.; Yardımcı, R.E.; Eryalçın, K.M. The effect of feeding with different microalgae on survival, growth, and fatty acid composition of *Artemia franciscana* metanauplii and on predominant bacterial species of the rearing water. *Aquacult. Int.* **2021**, *29*, 2223–2241. [CrossRef]
35. Sargent, J.; McEvoy, L.; Estevez, A.; Bell, G.; Bell, M.; Henderson, J.; Tocher, D. Lipid nutrition of marine fish during early development: Current status and future directions. *Aquaculture* **1999**, *179*, 217–229. [CrossRef]
36. Triantaphyllidis, G.V.; Abatzopoulos, T.J.; Sorgeloos, P. Review of the biogeography of the genus *Artemia* (Crustacea. Anostraca). *J. Biogeogr.* **1998**, *25*, 213–226. [CrossRef]
37. Maldonado-Montiel, T.D.N.J.; Rodriguez-Canche, L.G.; Olveranova, M.A. Evaluation of *Artemia* biomass production in San Crisanto, Yucatan, Mexico, with the use of poultry manure as organic fertilizer. *Aquaculture* **2003**, *219*, 573–584. [CrossRef]
38. Kokalj, A.J.; Kunej, U.; Skalar, T. Screening study of four environmentally relevant microplastic pollutants: Uptake and effects on *Daphnia magna* and *Artemia franciscana*. *Chemosphere* **2018**, *208*, 522–529. [CrossRef] [PubMed]
39. Li, H.; Chen, H.; Wang, J.; Li, J.; Liu, S.; Tu, J.; Chen, Y.; Zong, Y.; Zhang, P.; Wang, Z.; et al. Influence of Microplastics on the Growth and the Intestinal Microbiota Composition of Brine Shrimp. *Front. Microbiol.* **2021**, *12*, 717272. [CrossRef] [PubMed]
40. Batel, A.; Linti, F.; Scherer, M.; Erdinger, L.; Braunbeck, T. Transfer of benzo[a]pyrene from microplastics to *Artemia* nauplii and further to zebrafish via a trophic food web experiment: CYP1A induction and visual tracking of persistent organic pollutants. *Environ. Toxicol. Chem.* **2016**, *35*, 1656–1666. [CrossRef] [PubMed]
41. Nunes, B.S.; Carvalho, F.D.; Guilhermino, L.M.; Van Stappen, G. Use of the genus *Artemia* in ecotoxicity testing. *Environ. Pollut.* **2006**, *144*, 453–462. [CrossRef] [PubMed]
42. Libralato, G. The case of *Artemia* spp. in nanoecotoxicology. *Mar. Environ. Res.* **2014**, *101*, 38–43. [CrossRef]
43. Castritsi-Catharios, J.; Syriou, V.; Miliou, H.; Zouganelis, G.D. Toxicity effects of bisphenol A to the nauplii of the brine shrimp *Artemia* franciscana. *J. Biol. Res. Thessalon.* **2013**, *19*, 38–45.
44. Ekonomou, G.; Lolas, A.; Castritsi-Catharios, J.; Neofitou, C.; Zouganelis, G.D.; Tsiropoulos, N.; Exadactylos, A. Mortality and Effect on Growth of *Artemia franciscana* Exposed to Two Common Organic Pollutants. *Water* **2019**, *11*, 1614. [CrossRef]
45. Shaukat, A.; Liu, G.; Li, Z.; Xu, D.; Huang, T.; Chen, H. Toxicity of five phenolic compounds to brine shrimp *Artemia sinica* (Crustacea: Artemiidae). *J. Ocean Univ. China* **2014**, *13*, 141–145. [CrossRef]
46. Kidd, K.A.; Paterson, M.J.; Rennie, M.D.; Podemski, C.L.; Findlay, D.L.; Blanchfield, P.J.; Liber, K. Direct and indirect responses of a freshwater food web to a potent synthetic oestrogen. *Philos. Trans. R. Soc. B Biol. Sci.* **2014**, *369*, 20130578. [CrossRef]
47. Jeong, T.Y.; Simpson, M.J. Endocrine Disruptor Exposure Causes Infochemical Dysregulation and an Ecological Cascade from Zooplankton to Algae. *Environ. Sci. Technol.* **2021**, *55*, 3845–3854. [CrossRef] [PubMed]
48. Ballesteros-Gómez, A.; Rubio, S.; Pérez-Bendito, D. Analytical methods for the determination of bisphenol A in food. *J. Chromatogr. A* **2009**, *1216*, 449–469. [CrossRef] [PubMed]
49. Corrales, J.; Kristofco, L.A.; Steele, W.B.; Yates, B.S.; Breed, C.S.; Williams, E.S.; Brooks, B.W. Global Assessment of Bisphenol A in the Environment: Review and Analysis of Its Occurrence and Bioaccumulation. *Dose-Response* **2015**, *13*, 1–29. [CrossRef] [PubMed]
50. Xiao, Q.; Li, Y.; Ouyang, H.; Xu, P.; Wu, D. High-performance liquid chromatographic analysis of bisphenol A and 4-nonylphenol in serum, liver and testis tissues after oral administration to rats and its application to toxicokinetic study. *J. Chromatogr. B Anal. Technol. Biomed. Life Sci.* **2006**, *830*, 322–329. [CrossRef] [PubMed]
51. Samanidou, V.F.; Frysali, M.A.; Papadoyannis, I.N. Matrix solid phase dispersion for the extraction of bisphenol A from human breast milk prior to HPLC analysis. *J. Liq. Chromatogr. Relat. Technol.* **2014**, *37*, 247–258. [CrossRef]

52. Samanidou, V.F.; Kerezoudi, C.; Tolika, E.; Palaghias, G. A Simple Isocratic HPLC Method for the Simultaneous Determination of the Five Most Common Residual Monomers Released from Resin-Based Dental Restorative Materials. *J. Liq. Chromatogr. Relat. Technol.* **2015**, *38*, 740–749. [CrossRef]
53. Samanidou, V.; Hadjicharalampous, M.; Palaghias, G.; Papadoyannis, I. Development and validation of an isocratic HPLC method for the simultaneous determination of residual monomers released from dental polymeric materials in artificial saliva. *J. Liq. Chromatogr. Relat. Technol.* **2012**, *35*, 511–523. [CrossRef]
54. Diamantopoulou, E.I.; Plastiras, O.E.; Mourouzis, P.; Samanidou, V. Validation of a Simple HPLC-UV Method for the Determination of Monomers Released from Dental Resin Composites in Artificial Saliva. *Methods Protoc.* **2020**, *3*, 35. [CrossRef]
55. Liu, J.; Pan, X.; Huang, B.; Fang, K.; Wang, Y.; Gao, J. An improved method for simultaneous analysis of steroid and phenolic endocrine disrupting chemicals in biological samples. *J. Environ. Anal. Chem.* **2012**, *92*, 1135–1149. [CrossRef]
56. Asimakopoulos, A.G.; Thomaidis, N.S.; Koupparis, M.A. Recent trends in biomonitoring of bisphenol A, 4-t-octylphenol, and 4-nonylphenol. *Toxicol. Lett.* **2012**, *210*, 141–154. [CrossRef] [PubMed]
57. Preindl, K.; Braun, D.; Aichinger, G.; Sieri, S.; Fang, M.; Marko, D.; Warth, B. A Generic Liquid Chromatography-Tandem Mass Spectrometry Exposome Method for the Determination of Xenoestrogens in Biological Matrices. *Anal. Chem.* **2019**, *91*, 11334–11342. [CrossRef] [PubMed]
58. Aznar, R.; Albero, B.; Pérez, R.A.; Sánchez-Brunete, C.; Miguel, E.; Tadeo, J.L. Analysis of emerging organic contaminants in poultry manure by gas chromatography-tandem mass spectrometry. *J. Sep. Sci.* **2018**, *41*, 940–947. [CrossRef] [PubMed]
59. Staniszewska, M.; Koniecko, I.; Falkowska, L.; Krzymyk, E. Occurrence and distribution of bisphenol A and alkylphenols in the water of the gulf of Gdansk (southern Baltic). *Mar. Pollut. Bull.* **2015**, *91*, 372–379. [CrossRef]
60. Staniszewska, M.; Graca, B.; Sokołowski, A.; Nehring, I.; Wasik, A.; Jendzul, A. Factors determining accumulation of bisphenol A and alkylphenols at a low trophic level as exemplified by mussels *Mytilus trossulus*. *Environ. Pollut.* **2017**, *220*, 1147–1159. [CrossRef]
61. Staniszewska, M.; Nehring, I.; Mudrak- Cegiołka, S. Changes of concentrations and possibility of accumulation of bisphenol A and alkylphenols, depending on biomass and composition, in zooplankton of the Southern Baltic (Gulf of Gdansk). *Environ. Pollut.* **2016**, *10*, 489–501. [CrossRef]
62. Rigopoulos, A.T.; Samanidou, V.F.; Touraki, M. Development and Validation of an HPLC-DAD Method for the Simultaneous Extraction and Quantification of Bisphenol-A, 4-Hydroxybenzoic Acid, 4-Hydroxyacetophenone and Hydroquinone in Bacterial Cultures of Lactococcus lactis. *Separations* **2018**, *5*, 12. [CrossRef]
63. Samat, N.A.; Yusoff, F.M.; Rasdi, N.W.; Karim, M. Enhancement of Live Food Nutritional Status with Essential Nutrients for Improving Aquatic Animal Health: A Review. *Animals* **2020**, *10*, 2457. [CrossRef]
64. Varó, I.; Serrano, R.; Pitarch, E.; Amat, F.; López, F.J.; Navarro, J.C. Bioaccumulation of chlorpyrifos through an experimental food chain: Study of protein HSP70 as biomarker of sublethal stress in fish. *Arch. Environ. Contam. Toxicol.* **2002**, *42*, 229–235. [CrossRef]
65. Correa-Reyes, G.; Viana, M.T.; Marquez-Rocha, F.J.; Licea, A.F.; Ponce, E.; Vazquez-Duhalt, R. Nonylphenol algal bioaccumulation and its effect through the trophic chain. *Chemosphere* **2007**, *68*, 662–670. [CrossRef]
66. Dorn, P.B.; Chou, C.S.; Getempo, J.J. Degradation of Bisphenol A in Natural Waters. *Chemosphere* **1987**, *16*, 1501–1507. [CrossRef]
67. Chapin, R.E.; Adams, J.; Boekelheide, K.; Gray, L.E., Jr.; Hayward, S.W.; Lees, P.S.; McIntyre, B.S.; Portier, K.M.; Schnorr, T.M.; Selevan, S.G.; et al. NTP-CERHR expert panel report on the reproductive and developmental toxicity of bisphenol A. *Birth Defects Res. B Dev. Reprod. Toxicol.* **2008**, *83*, 157–395. [CrossRef] [PubMed]
68. EFSA. Scientific Opinion on the risks to public health related to the presence of bisphenol A (BPA) in foodstuffs. *EFSA J.* **2015**, *13*, 3978. [CrossRef]
69. National Center for Biotechnology Information. PubChem Compound Summary for CID 8814, 4-tert-Octylphenol. Available online: https://pubchem.ncbi.nlm.nih.gov/compound/4-tert-Octylphenol (accessed on 14 February 2022).
70. Jonsson, B. Risk Assessment on Butylphenol, Octylphenol and Nonylphenol, and Estimated Human Exposure of Alkylphenols from Swedish Fish. Uppsala Uppsala University. 2006. Available online: https://www.uu.se/digitalAssets/177/c_177024-l_3-k_jonsson-beatrice-report.pdf (accessed on 22 April 2022).
71. Profiles of the Initial Environmental Risk Assessment of Chemicals Vol. 7, Ministry of the Environment, Government. Available online: https://www.env.go.jp/en/chemi/chemicals/profile_erac/index.html#vol7 (accessed on 22 April 2022).
72. Touraki, M.; Karamanlidou, G.; Karavida, P.; Chrysi, K. Evaluation of the probiotics *Bacillus subtilis* and *Lactobacillus plantarum* bioencapsulated in *Artemia* nauplii against vibriosis in European sea bass larvae (*Dicentrarchus labrax*, L.). *World J. Microbiol. Biotechnol.* **2012**, *28*, 2425–2433. [CrossRef]
73. Giarma, E.; Amanetidou, E.; Toufexi, A.; Touraki, M. Defense systems in developing *Artemia franciscana* nauplii and their modulation by probiotic bacteria offer protection against a *Vibrio anguillarum* challenge. *Fish Shellfish Immunol.* **2017**, *66*, 163–172. [CrossRef] [PubMed]
74. Pedersen, S.N.; Lindholst, C. Quantification of the xenoestrogens 4-tert-octylphenol and bisphenol A in water and in fish tissue based on microwave assisted extraction, solid-phase extraction and liquid chromatography-mass spectrometry. *J. Chromatogr. A* **1999**, *864*, 17–24. [CrossRef]
75. Singh, B.; Kumar, A.; Malik, A.K. Recent advances in sample preparation methods for analysis of endocrine disruptors from various matrices. *Crit. Rev. Anal. Chem.* **2014**, *44*, 255–269. [CrossRef]

76. Cerkvenik-Flajs, V.; Fonda, I.; Gombač, M. Analysis and Occurrence of Bisphenol A in Mediterranean Mussels (*Mytilus galloprovincialis*) Sampled from the Slovenian Coastal Waters of the North Adriatic Sea. *Bull. Environ. Contam. Toxicol.* **2018**, *101*, 439–445. [CrossRef]
77. Völkel, W.; Colnot, T.; Csanady, G.; Filser, J.G.; Dekant, W. Metabolism and kinetics of bisphenol A in humans at low doses following oral administration. *Chem. Res. Toxicol.* **2002**, *15*, 1281–1287. [CrossRef]
78. Dong, C.D.; Chen, C.W.; Chen, C.F. Seasonal and spatial distribution of 4-nonylphenol and 4-tert-octylphenol in the sediment of Kaohsiung Harbor, Taiwan. *Chemosphere* **2015**, *134*, 588–597. [CrossRef]
79. Pisciottano, I.D.M.; Mita, G.D.; Gallo, P. Bisphenol A, octylphenols and nonylphenols in fish muscle determined by LC/ESI-MS/MS after affinity chromatography clean up. *Food Addit. Contam. Part B* **2020**, *13*, 139–147. [CrossRef] [PubMed]
80. Cañadas, R.; Garrido Gamarro, E.; Garcinuño Martínez, R.M.; Paniagua González, G.; Fernández Hernando, P. Occurrence of common plastic additives and contaminants in mussel samples: Validation of analytical method based on matrix solid-phase dispersion. *Food Chem.* **2021**, *349*, 129169. [CrossRef] [PubMed]
81. Basheer, C.; Lee, H.K.; Tan, K.S. Endocrine disrupting alkylphenols and bisphenol-A in coastal waters and supermarket seafood from Singapore. *Mar. Pollut. Bull.* **2004**, *48*, 1161–1167. [CrossRef] [PubMed]
82. Uhrovčík, J. Strategy for determination of LOD and LOQ values—Some basic aspects. *Talanta* **2014**, *119*, 178–180. [CrossRef] [PubMed]
83. McMillan, J. Principles of analytical validation. In *Proteomic Profiling and Analytical Chemistry: The Crossroads*, 2nd ed.; Ciborowski, P., Silberring, J., Eds.; Elsevier: Amsterdam, The Netherlands, 2016; pp. 205–215. [CrossRef]
84. Shaukat, A.; Guangxing, L.; Zhengyan, L. The Acute Toxicity of Phenolic Compounds and Heavy Metals on Brine Shrimp *Artemia sinica* (Crustacea: Artemiidae). *Sindh Univ. Res. J.* **2013**, *45*, 21–24.
85. Rajasree, S.R.R.; Kumar, V.G.; Abraham, L.S.; Manoharan, N. Assessment on the toxicity of engineered nanoparticles on the life stages of marine aquatic invertebrate *Artemia salina*. *Int. J. Nanosci.* **2011**, *10*, 1153–1159. [CrossRef]
86. Mtibaà, R.; Ezzanad, A.; Aranda, E.; Pozo, C.; Ghariani, B.; Moraga, J.; Nasri, M.; Manuel Cantoral, J.; Garrido, C.; Mechichi, T. Biodegradation and toxicity reduction of nonylphenol, 4-tert-octylphenol and 2,4-dichlorophenol by the ascomycetous fungus *Thielavia* sp HJ22: Identification of fungal metabolites and proposal of a putative pathway. *Sci. Total Environ.* **2020**, *708*, 135129. [CrossRef]
87. Olaniyan, L.W.B.; Okoh, A.I. Determination and ecological risk assessment of two endocrine disruptors from River Buffalo, South Africa. *Environ. Monit. Assess.* **2020**, *192*, 750. [CrossRef]
88. Gurban, A.M.; Burtan, D.; Rotariu, L.; Bala, C. Manganese oxide based screen-printed sensor for xenooestrogens detection. *Sens. Actuators B Chem.* **2015**, *210*, 273–280. [CrossRef]
89. Neamtu, M.; Frimmel, F.H. Degradation of endocrine disrupting bisphenol A by 254 nm irradiation in different water matrices and effect on yeast cells. *Water Res.* **2006**, *40*, 3745–3750. [CrossRef]
90. Wu, Y.; Yuan, H.; Wei, G.; Zhang, S.; Li, H.; Dong, W. Photodegradation of 4-tert octylphenol in aqueous solution promoted by Fe (III). *Environ. Sci. Pollut. Res. Int.* **2013**, *20*, 3–9. [CrossRef] [PubMed]
91. Ishihara, K.; Nakajima, N. Improvement of marine environmental pollution using eco-system: Decomposition and recovery of endocrine disrupting chemicals by marine phyto- and zooplanktons. *J. Mol. Catal. B* **2003**, *23*, 419–424. [CrossRef]
92. Ocharoen, Y.; Boonphakdee, C.; Boonphakdee, T.; Shinn, A.P.; Moonmangmee, S. High levels of the endocrine disruptors bisphenol-A and 17β-estradiol detected in populations of green mussel, Perna viridis, cultured in the Gulf of Thailand. *Aquaculture* **2018**, *497*, 348–356. [CrossRef]
93. Lu, I.-C.; Chao, H.-R.; Mansor, W.-N.-W.; Peng, C.-W.; Hsu, Y.-C.; Yu, T.-Y.; Chang, W.-H.; Fu, L.-M. Levels of Phthalates, Bisphenol-A, Nonylphenol, and Microplastics in Fish in the Estuaries of Northern Taiwan and the Impact on Human Health. *Toxics* **2021**, *9*, 246. [CrossRef]
94. Yap, C.K. Contamination of Heavy Metals and Other Organic Pollutants in Perna Viridis from the Coastal Waters of Malaysia: A Review Based On 1998 Data. *J. Sci. Res. Rep.* **2014**, *3*, 1–16. [CrossRef] [PubMed]
95. Arditsoglou, A.; Voutsa, D. Occurrence and partitioning of endocrine-disrupting compounds in the marine environment of Thermaikos Gulf, Northern Aegean Sea, Greece. *Mar. Pollut. Bull.* **2012**, *64*, 2443–2452. [CrossRef]
96. Giebichenstein, J.; Giebichenstein, J.; Hasler, M.; Schulz, C.; Ueberschär, B. Comparing the performance of four commercial microdiets in an early weaning protocol for European seabass larvae (*Dicentrarchus labrax*). *Aquac. Res.* **2022**, *53*, 544–558. [CrossRef]
97. Capó, X.; Alomar, C.; Compa, M.; Sole, M.; Sanahuja, I.; Soliz Rojas, D.L.; González, G.P.; Garcinuño Martínez, R.M.; Deudero, S. Quantification of differential tissue biomarker responses to microplastic ingestion and plasticizer bioaccumulation in aquaculture reared sea bream *Sparus aurata*. *Environ. Res.* **2022**, *211*, 113063. [CrossRef]

Article

# Allergens and Other Harmful Substances in Hydroalcoholic Gels: Compliance with Current Regulation

Ana Castiñeira-Landeira [1], Lua Vazquez [1], Thierry Dagnac [2], Maria Celeiro [1,*] and María Llompart [1,*]

1. CRETUS, Department of Analytical Chemistry, Nutrition and Food Science, Universidade de Santiago de Compostela, E-15782 Santiago de Compostela, Spain; anacastineira.landeira@usc.es (A.C.-L.); lua.vazquez.ferreiro@usc.es (L.V.)
2. Galician Agency for Food Quality, Agronomic Research Centre (AGACAL-CIAM), Unit of Organic Contaminants, P.O. Box 10, E-15080 A Coruña, Spain; thierry.dagnac@xunta.gal
* Correspondence: maria.celeiro.montero@usc.es (M.C.); maria.llompart@usc.es (M.L.); Tel.: +34-88181-4225 (M.L.)

**Citation:** Castiñeira-Landeira, A.; Vazquez, L.; Dagnac, T.; Celeiro, M.; Llompart, M. Allergens and Other Harmful Substances in Hydroalcoholic Gels: Compliance with Current Regulation. *Methods Protoc.* **2023**, *6*, 95. https://doi.org/10.3390/mps6050095

Academic Editors: Victoria Samanidou, Verónica Pino and Natasa Kalogiouri

Received: 31 July 2023
Revised: 28 September 2023
Accepted: 30 September 2023
Published: 7 October 2023

**Copyright:** © 2023 by the authors. Licensee MDPI, Basel, Switzerland. This article is an open access article distributed under the terms and conditions of the Creative Commons Attribution (CC BY) license (https://creativecommons.org/licenses/by/4.0/).

**Abstract:** Hydroalcoholic gels or hand sanitisers have become essential products to prevent and mitigate the transmission of COVID-19. Depending on their use, they can be classified as cosmetics (cleaning the skin) or biocides (with antimicrobial effects). The aim of this work was to determine sixty personal care products frequently found in cosmetic formulations, including fragrance allergens, synthetic musks, preservatives and plasticisers, in hydroalcoholic gels and evaluate their compliance with the current regulation. A simple and fast analytical methodology based on solid-phase microextraction followed by gas chromatography–tandem mass spectrometry (SPME-GC-MS/MS) was validated and applied to 67 real samples. Among the 60 target compounds, 47 of them were found in the analysed hand sanitisers, highlighting the high number of fragrance allergens (up to 23) at concentrations of up to 32,458 µg g$^{-1}$. Most of the samples did not comply with the labelling requirements of the EU Regulation No 1223/2009, and some of them even contained compounds banned in cosmetic products such as plasticisers. Method sustainability was also evaluated using the metric tool AGREEPrep, demonstrating its greenness.

**Keywords:** hydroalcoholic gels; personal care products; solid phase microextraction; gas chromatography; tandem mass spectrometry

## 1. Introduction

Due to the COVID-19 (severe acute respiratory syndrome coronavirus 2, SARS-CoV-2) infectious disease that was declared a pandemic by the World Health Organization (WHO) in March 2020, promoting hand hygiene by means of alcohol-based hand rub (ABHR) has been considered the primary strategy to mitigate its transmission and infection [1]. In this sense, the production and consumption of hand sanitisers drastically increased worldwide (by up to 561% in Italy, one of the most affected European countries) [2,3]. These products were massively placed in public areas such as shopping centres, schools, banks, supermarkets for public use. Hand sanitisers can be defined as 'borderline products' since their classification as a biocide or as a cosmetic product is not clear; this depends on the presence of an active substance and the product's main purpose [4]. Those designed to disinfect hands, eliminating the microorganisms and their possible transmission, are considered as biocides and are subject to Biocidal Products Regulation (EU) No 528/2012 [5]. On the other hand, if their main purpose is cleaning or cleansing the skin, notably in the absence of water rinsing, then they are considered cosmetics, and probably do not protect through biocidal action. In this case, they must comply with the Cosmetics Regulation (EU) No 1223/2009 [6].

To homogenise their formulation and fabrication, ensuring their antimicrobial properties in the context of the COVID-19 pandemic, the WHO published a protocol, and two formulations were established, containing: (i) ethanol (96%), hydrogen peroxide (3%) and glycerol (98%) and (ii) isopropyl alcohol (99.8%), hydrogen peroxide (3%) and glycerol (98%) [7]. This protocol also strongly recommended that no ingredients other than those specified above be added to the formulation, whilst the addition of fragrances was not recommended because of the risk of allergic reactions. Although it is well known that fragrances are the main responsible of allergic contact dermatitis (ACD), they are usually added to cosmetics with the intention of providing a pleasant scent. Other compounds such as preservatives and plasticisers can also be present in cosmetics and daily care products. Preservatives are added to protect products and consumers against microbial growth, whereas plasticisers (mainly phthalates and adipates) are added to fix and dissolve fragrances, although some can be transferred from the plastic containers to the product and then to consumers, causing health problems, since many of them are catalogued as endocrine disruptors [8–11].

Different studies have demonstrated that SARS-CoV-2 was efficiently inactivated by WHO-recommended formulations, reducing the viral load to a background level within 30 s, whereas formulations containing other additives present a lower effectivity against the virus [12–14].

Therefore, the main objective of this work was the validation and application of a sensitive analytical method to simultaneously determine fragrances (allergens and synthetic musks) and other potentially harmful substances, such as plasticisers and preservatives, including 60 compounds, in a broad range of hand sanitiser samples. In general, the analysis of cosmetics and personal care products is a challenge for analysts, as ingredients may be present at % concentrations, or at trace levels in the case of impurities. A suitable option for the multianalyte analysis of these matrices is the use of solid-phase microextraction (SPME), a simple, fast and environmentally friendly extraction technique, followed by gas chromatography–tandem mass spectrometry (GC-MS/MS). This combination has been reported as a reliable and sensitive tool to analyse a high number of personal care products (PCPs) in different cosmetic products, including hydroalcoholic gels [15–18]. The method was employed to conduct a survey of hand sanitisers for personal use (personal use) and placed in local shops or public areas (public use) to verify compliance with applicable legislation. A labelling study was also carried out to check whether the information for consumers was correct and complete. To the best of our knowledge, there are no analytical studies dedicated to researching this type of samples used daily by a large part of the world's population since the COVID-19 pandemic.

## 2. Materials and Methods

### 2.1. Chemicals, Reagents and Materials

The 60 target compounds, their CAS numbers, molecular mass, European legislation restrictions, retention times and MS/MS transitions are shown in Table S1.

Ultrapure water and methanol, both MS grade, were purchased from Scharlau (Barcelona, Spain) and acetone from Sigma Aldrich Chemie GmbH (Steinheim, Germany). Moreover, 50/30 μm of commercial divinylbenzene/carboxen/polydimethylsiloxane (DVB/CAR/PDMS) fibre housed in a manual SPME holder was obtained from Supelco (Bellefonte, PA, USA). Prior to the first use, the fibre was conditioned as recommended by the manufacturer by inserting it into the GC injector with carrier gas flow at 270 °C for 30 min.

The target compounds were selected, including 23 fragrance allergens, 11 synthetic musks, 16 plasticisers and 10 preservatives. Individual stock solutions were prepared in methanol or ethyl acetate, followed by further dilutions in acetone. Working solutions were prepared weekly. All of them were stored in amber glass vials and protected from light at −20 °C.

### 2.2. Samples and Sampling

Sixty-seven hydroalcoholic gel samples were collected from different establishments in Galicia (Northwest Spain), including banks, restaurants, pharmacies, universities, local markets in 15 mL amber vials. The samples were kept at room temperature and protected from light until analysis. Sampling details about the establishments where they were collected, and the composition indicated in the label, are summarised in Table S2.

### 2.3. Sample Preparation Procedure: SPME

Herein, 10 mg of hydroalcoholic gel and 10 mL of ultrapure water (1:1000, $w/v$ dilution) were placed in a 20 mL glass vial, which was sealed with an aluminium cap furnished with PTFE-faced septa. The vial with the sample was immersed in a water bath at 100 °C. First, the sample was conditioned during 5 min and then the DVB/CAR/PDMS fibre was exposed for 20 min to the headspace over the sample (HS-SPME mode). Agitation under magnetic stirring was employed during the extraction procedure. After the extraction time, the SPME fibre was desorbed at 270 °C for 5 min in the GC injection port and the GC-MS/MS analysis was carried out.

In some cases, the high concentration of some of the analytes in the samples made it necessary to perform a further dilution of the hydroalcoholic gel–water mixture in ultrapure water, followed by the corresponding SPME GC-MS/MS analysis.

Since one of the studied families are plasticisers, and to avoid contamination and overestimation in the results, the plastic material was replaced with glass and metallic. All material was also maintained at 230 °C for 12 h before use. In addition, fibre blanks and procedure blanks using 10 mL of ultrapure water were carried out with the aim of avoiding false-positive results.

### 2.4. GC-MS/MS Analysis

The GC-MS/MS analysis was performed employing a Thermo Scientific Trace 1310 gas chromatograph coupled to a triple-quadrupole mass spectrometer (TSQ 8000) from Thermo Scientific (San Jose, CA, USA).

Separation was performed on a Zebron ZB-Semivolatiles (30 m × 0.25 mm i.d. × 0.25 µm film thickness) obtained from Phenomenex (Torrance, CA, USA). Helium (purity 99.999%) was used as carrier gas at a constant column flow of 1 mL min$^{-1}$. The GC oven temperature was programmed from 60 °C (held 1 min), to 100 °C at 8 °C min$^{-1}$, to 150 °C at 20 °C min$^{-1}$, to 200 °C at 25 °C min$^{-1}$ (held 5 min), to 220 °C at 8 °C min$^{-1}$, and finally, to 290 °C at 30 °C min$^{-1}$ (held 7 min). The total run time was 30 min. Split/splitless mode was used for injection (200 kPa, held 1.2 min) and the injector temperature was kept at 270 °C.

The mass spectra detector (MSD) was operated in the electron impact (EI) ionisation positive mode (+70 eV). The temperatures of the transfer line and the ion source were established at 290 and 350 °C, respectively. The filament was set at 25 µA and the electron multiplier was set at a nominal value of 1800 V. Two or three transitions were monitored per compound working in the selected reaction monitoring (SRM) acquisition mode, as can be seen in Table S1. The system was operated by Xcalibur 2.2, and Trace Finder$^{TM}$ 3.2 software.

## 3. Results and Discussion

### 3.1. SPME-GC-MS/MS—Method Validation

The SPME experimental parameters, including fibre coating, extraction mode, temperature and dilution factor were previously optimised [17], and the most suitable conditions are described in Section 2.

The miniaturised method, employing only 10 mg of sample, was validated in terms of linearity, repeatability, accuracy and reproducibility. Instrumental detection limits (IDLs) and limits of detection (LODs) were also calculated. The results are summarised in Table S3. External calibration was assessed by preparing the standards in ultrapure water and following the procedure detailed in Section 2.3 covering a concentration range between 0.02 and 5 µg L$^{-1}$ (see specific ranges for each target analyte in Table S3). The method showed coefficients of determination ($R^2$) higher than 0.9901, demonstrating a direct proportional relationship between the amount of the target compound with the corresponding chromatographic response (area counts).

Precision was evaluated within a day ($n = 4$) and among several days ($n = 6$) for all the calibration levels. Table S3 shows relative standard deviation (RSD) values for 1 µg L$^{-1}$. As can be seen, they were lower than 13% and 16% for repeatability and reproducibility, respectively, in most cases. Accuracy was verified using an hydroalcoholic gel sample free of the target compounds which was fortified at two levels (0.2 µg g$^{-1}$ and 2 µg g$^{-1}$). Mean recoveries between 76 and 117% and precision (RSD%) < 10% were obtained, as can be seen in Figure S1.

IDLs were calculated as the compound concentration giving a signal-to-noise ratio of three (S/N = 3), employing ultrapure water standards containing low concentrations of the target analytes. LODs were calculated employing a real sample spiked with the target compounds. For the plasticisers that were detected in the procedure blanks (DEP, DBP, DEHP, DPhP, and DEHA), IDLs and LODs were calculated as the mean concentration corresponding to the signal of the blanks plus three times its standard deviation. Results are summarised in Table S3 and they were at the low ng g$^{-1}$ level for all compounds. In this case, it should be kept in mind that the gel samples are diluted by a factor of 1/1000, which explains the difference between IDL and LOD values. If higher sensitivity were required, a low dilution factor could be applied.

### 3.2. Greenness Assessment

In recent years, the development of extraction methodologies fulfilling green analytical chemistry (GAC) and green sample preparation (GSP) principles is increasing. These procedures include the use of safe solvents, reagents and materials minimising the experimental steps and reducing waste generation and energy consumption, allowing high sample throughput. In 2022, a metric tool, AGREEprep [19] was proposed for assessing the greenness of the sample preparation stage of an analytical procedure. The metric is based on ten principles of GSP, assigning weights at each criteria, depending on its significance in the analytical method. These weights range from 1, indicating low importance, to 5, representing high importance. Once the ten principles are evaluated, the tool is recalculated to the 0–1 scale to reflect the sample preparation greenness. The colour of the inner circle also provides the overall sample preparation greenness performance.

The sustainability of the SPME-GC-MS/MS method for extracting 60 personal care products from hydroalcoholic gels was calculated and results are depicted in Figure 1.

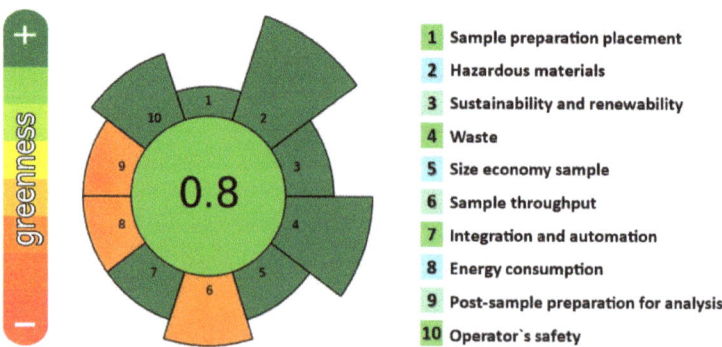

**Figure 1.** Evaluation of the degree of greenness. Pictogram obtained for SPME-GC-MS/MS under optimised conditions.

As can be seen, a value of 0.80 as well as a green label were obtained showing the greenness of the proposed SPME method. Considering each criteria, the developed method procedure is in-line/in situ since the SPME technique enables the integration of sample preparation and analysis (criteria 1); no toxic materials are used (criteria 2); the employed

SPME fibre can be used several times (criteria 3); no waste is generated because the samples are hydroalcoholic gels (criteria 4); the sample amount is 0.01 g (criteria 5); the duration of the sample preparation stage is around 20 min; three samples per hour (criteria 6); the procedure consists of two steps, extraction and desorption, and it is a fully automated system (criteria 7); the energy consumption is more than 183 Wh due to the heating magnetic stirrer (criteria 8); GC-MS/MS is used as analytical instrument (criteria 9); and, finally, no hazards are associated with the procedure (criteria 10). The weights of each criteria were, in general, not modified (default weights given by AGREEprep).

### 3.3. Analysis of Real Samples

The SPME-GC-MS/MS method was applied to analyse the target compounds in 67 hydroalcoholic gels, collected in different public places (see Section 2.2), in which 47 of the 60 target compounds were found (see Figure 2).

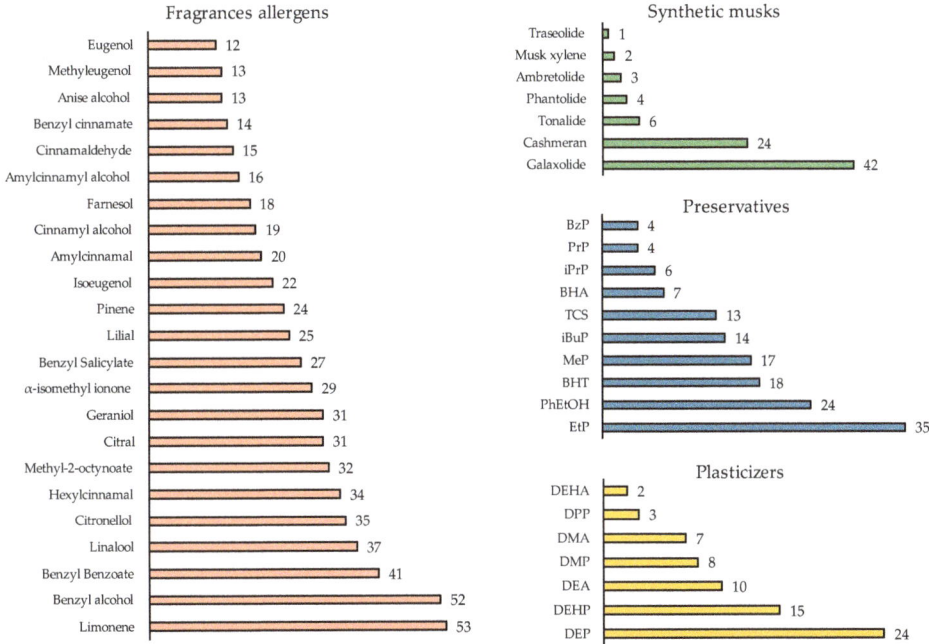

**Figure 2.** Frequency of the target compounds by families in the hydroalcoholic gels. Numbers indicate how many hydroalcoholic gels samples out of the 67 analysed contain each compound.

Most of the analysed samples do not comply with the WHO recommendations due to the presence of the target compounds (fragrances allergens, synthetic musks, plasticisers and preservatives) in the formulations. In this context, if the sample should be considered a cosmetic product, it must comply the European Cosmetics Regulation (EC) No 1223/2009 [6] in order to ensure product safety. This study highlights that 61% of the hydroalcoholic gels contain at least one compound which is prohibited by the Cosmetic Products Regulation, as can be seen in Figure 3.

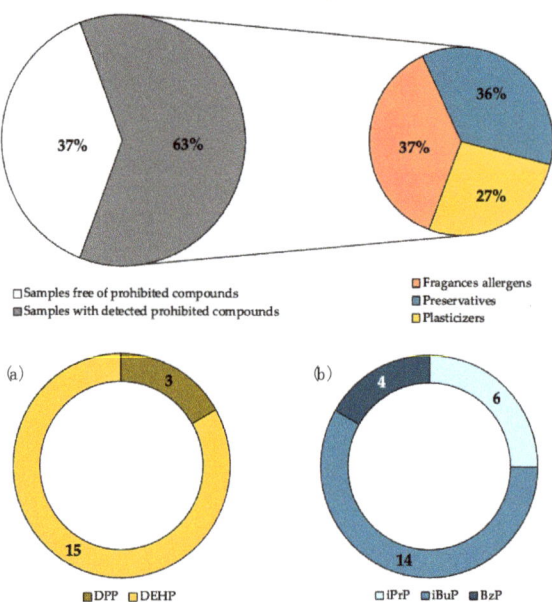

**Figure 3.** Percentage of banned compounds in the analysed samples and distribution by families: (**a**) phthalates and (**b**) parabens.

A description of the compliance with this regulation is presented below for each studied family. Table 1 summarises the concentration range, mean and median of the target compounds in the analysed samples, and the specific quantification results for the studied compounds in each sample are summarised in Table 2.

**Table 1.** Number of detected compounds, concentration range, mean and median, expressed at µg g$^{-1}$ in 67 analysed samples.

|  | Number | Range | Mean | Median |
|---|---|---|---|---|
| **Fragrance Allergens** | 65 | 0.0054–32,458 | 180 | 0.801 |
| Pinene | 24 | 0.022–4.4 | 1.0 | 0.59 |
| Limonene | 53 | 0.057–177 | 20 | 0.77 |
| Benzyl alcohol | 52 | 0.044–212 | 22 | 1.3 |
| Linalool | 37 | 0.036–209 | 35 | 4.9 |
| Methyl-2-octynoate | 32 | 0.072–72 | 7.1 | 1.25 |
| Citronellol | 35 | 0.0085–53 | 8.8 | 2.0 |
| Citral | 31 | 0.20–102 | 15 | 5.1 |
| Geraniol | 31 | 0.070–1747 | 103 | 12 |
| Cinnamaldehyde | 15 | 0.14–59 | 5.3 | 0.31 |
| Anise alcohol | 13 | 0.076–17,035 | 1375 | 11.7 |
| Cinnamyl alcohol | 19 | 0.17–327 | 51 | 3.7 |
| Eugenol | 12 | 0.026–110 | 23 | 6.5 |
| Methyl eugenol | 13 | 0.011–4.7 | 0.44 | 0.083 |
| Isoeugenol | 22 | 0.031–739 | 57 | 2.1 |
| α-isomethyl ionone | 29 | 0.0054–130 | 7.9 | 0.12 |
| Lilial | 25 | 0.0063–517 | 23 | 0.056 |
| Amyl cinnamal | 20 | 0.0094–32,458 | 1832 | 0.11 |
| Amylcinnamyl alcohol | 16 | 0.0072–427 | 45 | 0.58 |
| Farnesol | 18 | 0.0613–28,737 | 2363 | 7.07 |

Table 1. *Cont.*

|  | Number | Range | Mean | Median |
|---|---|---|---|---|
| **Fragrance Allergens** | 65 | 0.0054–32,458 | 180 | 0.801 |
| Hexyl cinnamal | 34 | 0.015–755 | 28 | 0.19 |
| Benzyl benzoate | 41 | 0.052–259 | 11 | 0.29 |
| Benzyl salicylate | 27 | 0.035–154 | 11 | 0.14 |
| Benzyl cinnamate | 14 | 0.019–0.84 | 0.15 | 0.058 |
| **Synthetic musks** | 51 | 0.0042–2356 | 56 | 0.16 |
| Musk xylene | 2 | 1840–2356 | 2098 | 2098 |
| Galaxolide | 42 | 0.019–56 | 5.8 | 0.17 |
| Phantolide | 4 | 0.011–29 | 7.3 | 0.081 |
| Cashmeran | 24 | 0.0042–74 | 4.2 | 0.16 |
| Traseolide | 1 | 1.8 | 1.8 | 1.8 |
| Tonalide | 6 | 0.022–0.386 | 0.131 | 0.0918 |
| Ambretolide | 3 | 0.0057–4.0 | 1.4 | 0.34 |
| **Preservatives** | 61 | 0.0035–13,735 | 365 | 1.59 |
| PhEtOH | 24 | 0.2121–13,735 | 2115 | 20.49 |
| BHA | 7 | 0.012–0.57 | 0.23 | 0.13 |
| BHT | 18 | 0.0035–170 | 10 | 0.043 |
| TCS | 13 | 0.039–0.45 | 0.15 | 0.10 |
| MeP | 17 | 0.18–23 | 4.4 | 1.3 |
| EtP | 35 | 0.58–62 | 9.0 | 3.6 |
| iPrP | 6 | 0.070–1.1 | 0.41 | 0.26 |
| PrP | 4 | 0.15–150 | 39 | 2.1 |
| iBuP | 14 | 0.15–61 | 14 | 4.1 |
| BzP | 4 | 3.0–59 | 21 | 11 |
| **Plasticisers** | 36 | 0.01–8238 | 256 | 0.54 |
| DMP | 8 | 0.16–108 | 15 | 0.56 |
| DEP | 24 | 0.19–8238 | 731 | 3.9 |
| DPP | 3 | 0.050–0.14 | 0.10 | 0.11 |
| DEHP | 15 | 0.11–12 | 1.4 | 0.25 |
| DMA | 7 | 0.11–4.07 | 0.87 | 0.23 |
| DEA | 10 | 0.010–1.8 | 0.48 | 0.11 |
| DEHA | 2 | 0.40–0.92 | 0.66 | 0.66 |

Table 2. Individual concentration (μg g$^{-1}$) of the fragrance allergens detected in 67 hydroalcoholic gel samples and total content.

| | Pinene | Limonene | Benzyl Alcohol | Linalool | Methyl-2-Octynoate | Citronellol | Citral | Geraniol | Cinnamal-dehyde | Anise Alcohol | Cinnamyl Alcohol | Eugenol |
|---|---|---|---|---|---|---|---|---|---|---|---|---|
| G1 | 0.022 ± 0.001 | | | 0.18 ± 0.06 | 0.14 ± 0.05 | 2.2 ± 0.6 | 0.4 ± 0.1 | 7 ± 1 | 59 ± 14 | | 71 ± 21 | |
| G2 | | 1.2 ± 0.1 | 2.3 ± 0.2 | 143 ± 5 | 0.27 ± 0.04 | 0.179 ± 0.001 | 2.7 ± 0.4 | 3.2 ± 0.3 | | | 1.0 ± 0.3 | |
| G3 | | 0.11 ± 0.01 | 0.171 ± 0.002 | 1.66 ± 0.01 | 0.41 ± 0.01 | 0.12 ± 0.02 | 0.21 ± 0.01 | | | | | |
| G4 | | 3.0 ± 0.7 | | | 2.1 ± 0.5 | | | | | | | |
| G5 | 0.89 ± 0.01 | 6.8 ± 0.5 | 10.2 ± 0.87 | 15 ± 1 | | 53 ± 3 | 98 ± 2 | 172 ± 12 | 0.227 ± 0.001 | | | 26 ± 3 |
| G6 | | 0.66 ± 0.03 | 0.67 ± 0.01 | 1.4 ± 0.016 | 0.65 ± 0.01 | | | | | 11 ± 1 | 159 ± 14 | |
| G7 | 0.19 ± 0.01 | 12 ± 1 | 11.7 ± 0.8 | 1.75 ± 0.08 | 2.58 ± 0.02 | 7.6 ± 0.9 | 36 ± 4 | 42 ± 5 | 0.17 ± 0.02 | | | 0.19 ± 0.02 |
| G8 | 0.08 ± 0.01 | 4 ± 1 | 4.5 ± 1.6 | | 1.37 ± 0.01 | 6.8 ± 0.1 | 3.8 ± 0.4 | 3.6 ± 0.2 | 0.14 ± 0.01 | | 2.5 ± 0.1 | |
| G9 | 1.0 ± 0.2 | 29 ± 9 | 28 ± 7 | 15 ± 4 | 4 ± 1 | | 25 ± 7 | 217 ± 44 | 5.6 ± 0.5 | | | 6 ± 2 |
| G10 | | 0.7 ± 0.1 | 0.9 ± 0.1 | 36.2 ± 0.4 | 2.96 ± 0.04 | 29 ± 0.5 | 2.9 ± 0.1 | 26 ± 1 | | | | 0.63 ± 0.08 |
| G11 | | 0.25 ± 0.05 | 0.22 ± 0.02 | 16 ± 2 | | | | | 0.18 ± 0.01 | | 0.46 ± 0.08 | |
| G12 | | | | | | | | | | | | |
| G13 | 0.83 ± 0.05 | 112 ± 4 | 121 ± 3 | 75 ± 2 | 40 ± 1 | 15 ± 3 | 41 ± 1 | 41 ± 1 | 0.21 ± 0.01 | | | |
| G14 | | 0.54 ± 0.01 | 0.47 ± 0.05 | 0.18 ± 0.04 | | | | | | | | |
| G15 | 0.27 ± 0.06 | 36 ± 11 | 37 ± 7 | 1.61 ± 0.09 | 0.19 ± 0.01 | | 2.1 ± 0.5 | | | | | |
| G16 | | | | | | 0.72 ± 0.02 | 0.27 ± 0.02 | | | | | |
| G17 | 0.3 ± 0.1 | 29 ± 8 | 26 ± 5 | 50 ± 10 | 3.2 ± 0.4 | 1.4 ± 0.2 | 5.6 ± 0.6 | 50 ± 7 | | | | |
| G18 | 1.94 ± 0.01 | 9.3 ± 1.2 | 9 ± 1 | 178 ± 2 | 12.2 ± 0.1 | | | 603 ± 15 | 7.20 ± 0.04 | 489 ± 14 | | |
| G19 | | 0.61 ± 0.04 | 0.59 ± 0.08 | 0.27 ± 0.04 | | | | | | 3.7 ± 0.2 | | |
| G20 | 0.97 ± 0.03 | 160 ± 2 | 181 ± 3 | 4.9 ± 0.2 | 0.82 ± 0.08 | | | 12.3 ± 0.3 | | | | |
| G21 | | 0.14 ± 0.04 | 0.17 ± 0.02 | | 1.6 ± 0.2 | | | | | | | |
| G22 | | 0.60 ± 0.06 | 0.300 ± 0.008 | | | 5.0 ± 0.3 | | | | | | |
| G23 | 4.3 ± 0.9 | 44 ± 7 | 57 ± 7 | | 72 ± 11 | 0.12 ± 0.02 | | 6 ± 1 | | | 13 ± 2 | |
| G24 | | 38 ± 3 | 40 ± 3 | 42 ± 5 | 45 ± 4 | 13 ± 1 | 102 ± 5 | 27 ± 5 | 1.6 ± 0.1 | | | |
| G25 | | | | | | 0.008 ± 0.001 | | | 0.30 ± 0.03 | | | |
| G26 | | | | | | | | | | | | |
| G27 | | | | | | | | | | | | |
| G28 | 2.7 ± 0.4 | 9 ± 1 | 10 ± 1 | | 1.1 ± 0.1 | 0.09 ± 0.01 | 1.1 ± 0.1 | 10 ± 2 | | | 5.5 ± 0.1 | |
| G29 | | 25 ± 2 | 23 ± 1 | 209 ± 17 | 3.31 ± 0.02 | 1.04 ± 0.02 | 7.27 ± 0.07 | | | | 0.6 ± 0.1 | |
| G30 | 3.6 ± 0.7 | 14 ± 1 | 13 ± 1 | 185 ± 25 | 13 ± 2 | | 17 ± 1 | 1747 ± 145 | | 17,036 ± 1039 | 327 ± 39 | |

Table 2. *Cont.*

| | Pinene | Limonene | Benzyl Alcohol | Linalool | Methyl-2-Octynoate | Citronellol | Citral | Geraniol | Cinnamal-dehyde | Anise Alcohol | Cinnamyl Alcohol | Eugenol |
|---|---|---|---|---|---|---|---|---|---|---|---|---|
| G31 | | | | | | 1.19 ± 0.06 | | | | | | |
| G32 | | | 58 ± 3 | 115 ± 17 | | | | | | | | |
| G33 | 0.438 ± 0.063 | 23.3 ± 3.3 | 25 ± 3 | 52 ± 3 | 0.791 ± 0.007 | 38 ± 3 | 12 ± 0.4 | 18 ± 3 | | | | 7.1 ± 0.2 |
| G34 | | | | 0.0363 ± 0.0025 | | | 0.29 ± 0.04 | 0.069 ± 0.002 | 0.90 ± 0.01 | 0.076 ± 0.002 | | |
| G35 | | 0.228 ± 0.072 | 0.179 ± 0.016 | | | | | | | | | |
| G36 | 2.7 ± 0.19 | 176 ± 7 | 212 ± 20 | 86 ± 5 | 1.768 ± 0.005 | 0.92 ± 0.06 | 37.8 ± 0.5 | 20.8 ± 0.79 | 2.5 ± 0.1 | 20.00 ± 0.02 | | 20.0 ± 0.6 |
| G37 | | 0.07 ± 0.01 | 0.057 ± 0.006 | | | | | | | | | |
| G38 | 0.28 ± 0.04 | 7 ± 2 | 8 ± 2 | 2.98 ± 0.01 | 0.135 ± 0.001 | 0.593 ± 0.008 | 5.0 ± 0.3 | 7.0 ± 0.1 | | | 3.7 ± 0.1 | |
| G39 | | 1.0 ± 0.1 | 0.54 ± 0.04 | | | | | | | | | 0.030 ± 0.007 |
| G40 | | 0.18 ± 0.024 | 0.10 ± 0.016 | | | | | | | | | |
| G41 | 2.3 ± 0.3 | 144 ± 21 | 167 ± 24 | 3.6 ± 0.3 | 0.51 ± 0.04 | 10.6 ± 0.6 | 7.7 ± 0.4 | 7.7 ± 0.53 | 0.17 ± 0.02 | | | 0.26 ± 0.013 |
| G42 | 0.7 ± 0.23 | 4.2 ± 0.3 | 3.8 ± 0.3 | 0.15 ± 0.01 | | 2.73 ± 0.10 | 0.37 ± 0.02 | | | | | |
| G43 | 0.032 ± 0.004 | 0.4 ± 0.1 | 0.42 ± 0.08 | 1.12 ± 0.03 | | 0.51 ± 0.02 | 0.54 ± 0.01 | 1.02 ± 0.01 | | 0.7 ± 0.2 | 1.51 ± 0.08 | |
| G44 | | 0.7 ± 0.23 | 0.7 ± 0.2 | | | | | | | | | |
| G45 | | 0.28 ± 0.01 | 0.22 ± 0.02 | | | | 0.20 ± 0.03 | | | | | |
| G46 | | 0.07 ± 0.02 | 0.05 ± 0.02 | | | | | | | | | |
| G47 | | | | | | | | | | | | |
| G48 | | 0.32 ± 0.10 | 0.31 ± 0.02 | 0.04 ± 0.01 | | 0.01 ± 0.01 | 0.22 ± 0.02 | | | | | |
| G49 | | | | | | | 0.20 ± 0.01 | | | | | |
| G50 | | 0.44 ± 0.02 | 1.5 ± 0.19 | 11 ± 1 | 0.07 ± 0.01 | 42 ± 2 | 16.4 ± 0.7 | 15.9 ± 0.74 | | 25.1 ± 1.5 | 2.0 ± 0.17 | 21 ± 1. |
| G51 | | 0.48 ± 0.08 | 5.0 ± 0.7 | 9.5 ± 1.8 | | 42 ± 7 | 26.2 ± 3.7 | 25 ± 3 | | 259 ± 32 | 16 ± 1 | 82 ± 6 |
| G52 | 0.14 ± 0.01 | 0.239 ± 0.06 | 0.26 ± 0.08 | | | 4.3 ± 1.1 | | 0.325 ± 0.051 | | | | |
| G53 | | 0.42 ± 0.09 | 0.48 ± 0.04 | 14.4 ± 1.8 | 0.74 ± 0.03 | 16.5 ± 2.1 | 8.725 ± 0.041 | 37.6 ± 4.1 | | 3.7 ± 1.1 | 156 ± 20 | 110 ± 28 |
| G54 | | | | | | | | | | | | |
| G55 | | 0.44 ± 0.02 | 0.24 ± 0.02 | | | 3.06 ± 0.17 | | | | | | |
| G56 | | 0.056 ± 0.008 | 0.057 ± 0.003 | | | 0.87 ± 0.13 | | | 0.658 ± 0.045 | | | |
| G57 | 0.130 ± 0.023 | 0.218 ± 0.0010 | 0.12 ± 0.02 | | | | | | | 0.82 ± 0.12 | 0.28 ± 0.04 | 0.02 ± 0.01 |
| G58 | | 0.067 ± 0.0053 | 0.093 ± 0.0072 | 0.177 ± 0.001 | | | | | | | | |
| G59 | | | | | 0.31 ± 0.07 | | | 1.71 ± 0.50 | | | | |
| G60 | 0.117 ± 0.03 | 19 ± 6 | 27 ± 8 | 0.16 ± 0.04 | 0.22 ± 0.08 | | | 0.13 ± 0.02 | | | 0.3 ± 0.1 | |

**Table 2.** *Cont.*

| | Pinene | Limonene | Benzyl Alcohol | Linalool | Methyl-2-Octynoate | Citronellol | Citral | Geraniol | Cinnamaldehyde | Anise Alcohol | Cinnamyl Alcohol | Eugenol |
|---|---|---|---|---|---|---|---|---|---|---|---|---|
| G61 | | | | | | | | | | | | |
| G62 | 0.85 ± 0.07 | 74.7 ± 0.8 | 73.78 ± 0.29 | 17.9 ± 1.6 | | 4.3 ± 0.4 | 12.2 ± 0.3 | 87 ± 11 | 0.23 ± 0.03 | | 200 ± 5 | |
| G63 | | 0.12 ± 0.01 | 0.08748 ± 0.00040 | 0.096 ± 0.008 | | 1.94 ± 0.017 | | | | | | |
| G64 | | | | 0.19 ± 0.03 | 2.423 ± 0.001 | 0.029 ± 0.007 | | | | | | |
| G65 | | 0.24 ± 0.02 | 2.9 ± 0.4 | | 12.6 ± 0.7 | 0.97 ± 0.01 | 0.9 ± 0.12 | 0.75 ± 0.08 | | 22 ± 8 | 18 ± 3 | |
| G66 | | 0.069 ± 0.003 | 0.04 ± 0.004 | | 0.188 ± 0.001 | 1.72 ± 0.04 | 0.20 ± 0.02 | 0.15 ± 0.04 | | 12 ± 2 | 0.17 ± 0.05 | |
| G67 | 0.095 ± 0.0098 | 1.2 ± 0.1 | 1.0 ± 0.07 | 0.10 ± 0.01 | 0.08 ± 0.01 | | | | | | | |

| | Methyl Eugenol | Isoeugenol | α-Isomethyl ionone | Lilial | Amyl Cinnamal | Amylcinnamyl Alcohol | Farnesol | Hexyl Cinnamal | Benzyl Benzoate | Benzyl Salicylate | Benzyl Cinnamate | Total Content |
|---|---|---|---|---|---|---|---|---|---|---|---|---|
| G1 | | 2.2 ± 0.5 | 2.5 ± 0.6 | 0.17 ± 0.045 | | 0.89 ± 0.27 | 9 ± 2 | | 18 ± 56 | 0.9 ± 0.2 | 0.3 ± 0.10 | 173 |
| G2 | | | 0.03 ± 0.01 | 0.0119 ± 0.001 | | | | | 0.5 ± 0.1 | 0.04 ± 0.01 | 0.020 ± 0.003 | 156 |
| G3 | | | | | | | | | 0.30 ± 0.01 | 0.034 ± 0.002 | 0.019 ± 0.001 | 3.06 |
| G4 | | 2.37 ± 0.05 | 0.44 ± 0.07 | | | 11 ± 1 | | | 0.7 ± 0.2 | | | 20.2 |
| G5 | 4.7 ± 0.5 | 1.17 ± 0.04 | 0.9 ± 0.1 | | | | | | 0.19 ± 0.01 | | | 391 |
| G6 | | | 0.045 ± 0.004 | 0.45 ± 0.02 | | | | 0.12 ± 0.01 | 0.23 ± 0.03 | 0.077 ± 0.002 | | 174 |
| G7 | | | 2.2 ± 0.3 | 4.91 ± 0.02 | | | 2.78 ± 0.02 | 1.1 ± 0.1 | 0.65 ± 0.05 | | | 126 |
| G8 | | | | | | | | 1.03 ± 0.09 | | | | 28.4 |
| G9 | | 3.6 ± 0.8 | 1.7 ± 0.3 | 6.0 ± 0.4 | | | | | 22 ± 5 | 4 ± 1 | | 369 |
| G10 | | 1.43 ± 0.09 | 9.95 ± 0.05 | 7.12 ± 0.02 | | 0.53 ± 0.09 | 10.8 ± 0.3 | | 9.9 ± 0.5 | 5.0 ± 0.8 | | 146 |
| G11 | | | | | | | | | | | | 17.1 |
| G12 | | | | | | | | | | | | 0 |
| G13 | | | | | | | | | 0.09 ± 0.03 | | | 448 |
| G14 | | | | | | | | | | | | 1.17 |
| G15 | | | | | | | | 1.4 ± 0.2 | 0.74 ± 0.02 | | | 80.2 |
| G16 | | | | | | | | | | | | 0.725 |
| G17 | | 7.8 ± 0.3 | | | 4180 ± 438 | | 32 ± 4 | 149 ± 26 | 31 ± 4 | 0.13 ± 0.01 | | 4567 |
| G18 | 0.05 ± 0.01 | | 0.06 ± 0.01 | | 0.74 ± 0.11 | | | 0.6 ± 0.3 | | | | 1311 |
| G19 | | 4.0 ± 0.1 | 0.135 ± 0.004 | 0.0066 ± 0.0004 | | | | 0.4 ± 0.2 | 64 ± 1 | | | 74.7 |
| G20 | | 0.208 ± 0.010 | | 517 ± 53 | | 0.290 ± 0.009 | | | 0.28 ± 0.02 | 5.5 ± 0.9 | 0.21 ± 0.01 | 884 |
| G21 | 0.06 ± 0.02 | | 2.3 ± 0.6 | | 0.33 ± 0.05 | 77.92 ± 0.07 | 4990 ± 5 | 0.14 ± 0.04 | | 0.11 ± 0.06 | 0.10 ± 0.03 | 5073 |

Table 2. Cont.

| | Methyl Eugenol | Isoeugenol | α-Isomethyl ionone | Lilial | Amyl Cinnamal | Amylcinna-myl Alcohol | Farnesol | Hexyl Cinnamal | Benzyl Benzoate | Benzyl Salicylate | Benzyl Cinnamate | Total Content |
|---|---|---|---|---|---|---|---|---|---|---|---|---|
| G22 | | | | | | | | | | | | 5.90 |
| G23 | 0.08 ± 0.01 | | | | 0.0341 ± 0.0082 | | 5.25 ± 0.85 | 0.01 ± 0.03 | | 0.28 ± 0.09 | | 202 |
| G24 | 0.092 ± 0.002 | | 0.06 ± 0.01 | 15 ± 1 | | 127 ± 1 | 28,737 ± 1 | | 0.27 ± 0.05 | 0.06 ± 0.02 | 0.06 ± 0.015 | 29,190 |
| G25 | | | | 0.08 ± 0.01 | | | | | | | | 0.640 |
| G26 | | | | | | | | 0.18 ± 0.02 | 0.115 ± 0.008 | 0.134 ± 0.005 | | 0.263 |
| G27 | | | | | | | | 0.076 ± 0.005 | 0.052 ± 0.007 | 0.07 ± 0.01 | 0.11 ± 0.01 | 0.187 |
| G28 | 0.09 ± 0.014 | 108 ± 1 | | | | | | | | | | 149 |
| G29 | 0.115 ± 0.005 | 1.004 ± 0.001 | 1.00 ± 0.03 | 10.12 ± 0.05 | | | | 0.292 ± 0.006 | 0.29 ± 0.01 | | | 292 |
| G30 | 0.047 ± 0.005 | 0.110 ± 0.005 | 0.11 ± 0.01 | | | | | 0.053 ± 0.009 | 9.912 ± 0.008 | 0.064 ± 0.004 | 0.050 ± 0.001 | 19,358 |
| G31 | | 0.65 ± 0.04 | | | | | | | | | | 1.83 |
| G32 | | | | | | | | | | | | 238 |
| G33 | | | 0.45 ± 0.03 | | 0.05 ± 0.00 | 0.6 ± 0.1 | | | | 0.345 ± 0.002 | | 256 |
| G34 | | 1.3 ± 0.19 | 0.005 ± 0.001 | | | | 1.8 ± 0.3 | 0.019 ± 0.001 | 0.055 ± 0.001 | | | 4.06 |
| G35 | | 8.01 ± 0.013 | | | | | | 0.12 ± 0.01 | 0.06 ± 0.019 | | | 8.60 |
| G36 | 0.20 ± 0.02 | 739 ± 7 | 0.018 ± 0.003 | | 0.115 ± 0.002 | 0.036 ± 0.001 | 1427 ± 21 | 0.52 ± 0.03 | 0.22 ± 0.01 | | | 2748 |
| G37 | | | | | | | | | | | | 0.135 |
| G38 | | 0.55 ± 0.01 | 0.102 ± 0.001 | 9.03 ± 0.03 | 0.106 ± 0.003 | | | 0.016 ± 0.003 | 0.053 ± 0.008 | | | 40.8 |
| G39 | | 0.80 ± 0.02 | | | | | 0.71 ± 0.08 | 0.016 ± 0.01 | | | | 6.95 |
| G40 | | | | | | | | | | | | 0.306 |
| G41 | 0.011 ± 0.003 | | | | | 0.087 ± 0.001 | | 6.3 ± 0.2 | 1.6 ± 0.29 | | | 353 |
| G42 | | | 0.008 ± 0.001 | 0.0252 ± 0.00067 | 0.059 ± 0.002 | | | 0.382 ± 0.005 | 0.49 ± 0.03 | 0.142 ± 0.005 | | 13.2 |
| G43 | | | 0.07 ± 0.01 | 0.018 ± 0.005 | 0.67 ± 0.05 | 0.193 ± 0.003 | 1.40 ± 0.03 | 0.70 ± 0.05 | 0.55 ± 0.04 | | 0.0268 ± 0.0076 | 10.0 |
| G44 | | | | | 0.01 ± 0.00 | | | 0.014 ± 0.001 | 0.132 ± 0.001 | | | 1.63 |
| G45 | | | | 0.0068 ± 0.00042 | 0.027 ± 0.005 | | | 0.24 ± 0.01 | 0.40 ± 0.01 | | | 1.39 |
| G46 | | | | | 0.011 ± 0.001 | | | 0.027 ± 0.004 | 0.13 ± 0.064 | | | 0.296 |
| G47 | | | | | | | | | | | | 0 |
| G48 | | | 0.009 ± 0.002 | 0.05 ± 0.010 | 0.009 ± 0.002 | | | 0.02 ± 0.01 | 0.23 ± 0.04 | | | 1.32 |
| G49 | | | | | | | | | 0.17 ± 0.03 | | 0.13 ± 0.022 | 0.580 |
| G50 | | 2.0 ± 1.1 | 3.45 ± 0.44 | 0.0212 ± 0.0072 | 0.740 ± 0.065 | 0.082 ± 0.044 | | 2.743 ± 0.003 | 1.905 ± 0.051 | 0.62 ± 0.07 | 0.05 ± 0.01 | 147 |
| G51 | 0.0227 ± 0.0015 | 23.5 ± 1.9 | 65.4 ± 7.2 | 0.082 ± 0.011 | 32,458 ± 41 | 2.71 ± 0.87 | | 755 ± 75 | | 154 ± 19 | 0.836 ± 0.004 | 33,925 |

Table 2. Cont.

| | Methyl Eugenol | Isoeugenol | α-Isomethyl ionone | Lilial | Amyl Cinnamal | Amylcinnamyl Alcohol | Farnesol | Hexyl Cinnamal | Benzyl Benzoate | Benzyl Salicylate | Benzyl Cinnamate | Total Content |
|---|---|---|---|---|---|---|---|---|---|---|---|---|
| G52 | | | | | | | | | | | | 5.23 |
| G53 | 0.123 ± 0.011 | | 130 ± 1 | 0.008 ± 0.001 | 1.1 ± 0.1 | 85 ± 16 | 6010 ± 163 | 0.23 ± 0.06 | 0.32 ± 0.02 | 0.38 ± 0.12 | | 6573 |
| G54 | | | | | | | 0.47 ± 0.02 | | 1.2 ± 0.37 | 3.3 ± 1.9 | | 8.63 |
| G55 | | | | | | | | 0.15 ± 0.02 | 0.08 ± 0.01 | 0.070 ± 0.006 | | 4.05 |
| G56 | 0.014 ± 0.005 | 0.031 ± 0.001 | | 0.02 ± 0.01 | 0.10 ± 0.02 | | | 0.10 ± 0.02 | 0.16 ± 0.01 | 0.139 ± 0.01 | | 2.21 |
| G57 | | | 0.035 ± 0.001 | | | | | | | | | 1.63 |
| G58 | | | | 0.019 ± 0.008 | | | | | | | | 0.358 |
| G59 | | | | | | | 34 ± 13 | | | | | 36.1 |
| G60 | | | | | | | | | | | | 48.0 |
| G61 | | | 0.012 ± 0.002 | | | | | | | 0.040 ± 0.004 | | 0.0523 |
| G62 | | 135 ± 3 | | 0.03 ± 0.001 | 0.039 ± 0.003 | 427 ± 4 | | | 259 ± 1 | 0.08 ± 0.01 | | 1293 |
| G63 | | | | 0.08 ± 0.01 | | | | 0.175 ± 0.006 | 0.11 ± 0.02 | 0.11 ± 0.01 | | 2.75 |
| G64 | | | | 0.032 ± 0.001 | | | | | | | | 2.68 |
| G65 | | 141 ± 11 | 0.029 ± 0.001 | | 0.18 ± 0.04 | 0.75 ± 0.06 | 1274 ± 85 | 0.116 ± 0.009 | 0.10 ± 0.01 | 27.0 ± 0.01 | 0.028 ± 0.007 | 1502 |
| G66 | | 80 ± 9 | 0.014 ± 0.002 | | 0.15 ± 0.07 | 0.007 ± 0.003 | 3.21 ± 0.01 | 0.19 ± 0.094 | 0.19 ± 0.056 | 93.44 ± 0.03 | 0.04 ± 0.02 | 191 |
| G67 | | | | 0.036 ± 0.005 | | | 0.9 ± 0.3 | | | | | 3.39 |

### 3.3.1. Fragrance Allergens

As can be seen in Tables 1 and 2, all studied fragrance allergens were detected in the real analysed samples. Limonene, benzyl alcohol and benzyl benzoate were found in most samples in 53, 52 and 41 out of the 67 hand sanitisers, respectively. Other fragrances such as linalool (37/67), citronellol (35/67), hexyl cinnamal (34/67), methyl-2-octynoate (32/67), citral and geraniol (31/67) were found in 50% of the samples. The presence of this family of cosmetic ingredients is remarkable since even the least frequently detected fragrances such as eugenol, anise alcohol and methyl eugenol, were found in 20% of the samples (see Figure 2).

Regarding the number of compounds per sample, 7 of the analysed samples contained around 17–19 of the 23 target fragrance allergens (G36, G43, G50, G51, G53, G65 and G66). In contrast, only two were free of these substances, samples G12 and G47, and only one sample (G16) contained one fragrance allergen (citronellol).

It should be underlined that the high levels found for some allergens in some samples reach concentrations of parts per hundred (up to 3%) as amyl cinnamal (32,458 µg g$^{-1}$ in sample G51) and farnesol (28,737 µg g$^{-1}$ in sample G24).

Regulatory Issues

According to EC Cosmetic Regulation No. 1223/2009, fragrance allergens can cause allergic skin reactions and other adverse effects especially at high concentrations. Among the 23 studied fragrance allergens, 20 (limonene, benzyl alcohol, linalool, methyl-2-octynoate, citronellol, citral, geraniol, cinnamaldehyde, anise alcohol, cinnamyl alcohol, eugenol, isoeugenol, α-isomethylionone, amyl cinnamal, amylcinnamyl alcohol, farnesol, hexyl cinnamal, benzyl salicylate and benzyl cinnamate) should appear on the label of cosmetic products when the concentration exceeds 0.001% ($w/w$, 10 µg g$^{-1}$) in leave-on products, as is the case for hand sanitisers [6]. Among the 67 analysed samples, 33 (49%) exceed this limit for several compounds; in addition, most of they are under-labelled since the corresponding fragrances are not included in the product label. In contrast, some hydroalcoholic gels were over-labelled, as they claimed to contain more allergens than they did (over-labelling is not included in the regulation). For example, sample G24 indicates the presence of benzyl benzoate, which is over-labelled since it does not surpass its legal limit (0.001%, 10 µg g$^{-1}$). In addition, two out of the studied fragrances present a maximum permitted concentration in final products as methyl-2-octynoate with 0.01% and isoeugenol with 0.02%. The sample G36 is the only one which surpassed the isoeugenol limit with a concentration of 739 µg g$^{-1}$.

Some fragrance allergens are prohibited, such as lilial, which has been banned since March 2022 because it damages fertility and it is suspected of damaging the unborn child, in addition to causing skin irritation [20]. Lilial was present in around the 37% (Figure 2) of the analysed hydroalcoholic gels (25/67) at concentrations below 0.001% ($w/w$, 10 µg g$^{-1}$) in 22 samples. Three samples (G10, G24 and G29) contained between 7 and 15 µg g$^{-1}$ of lilial, highlighting one sample which presented a very high value (517 µg g$^{-1}$, G20). However, it should be pointed out that some hydroalcoholic gel samples were taken after the lilial legislation had changed.

### 3.3.2. Synthetic Musks

Regarding the synthetic musks, as can be seen in Table 3, 7 of the 11 targets were found in the analysed hydroalcoholic gels, highlighting the presence of galaxolide in 63% of the samples (42/67), followed by cashmeran in 36% (24/67).

**Table 3.** Individual concentration (µg g$^{-1}$) of the synthetic musks detected in 67 hydroalcoholic gel samples and total content.

| | Musk Xylene | Galaxolide | Phantolide | Cashmeran | Traseolide | Tonalide | Ambrettolide | Total Content |
|---|---|---|---|---|---|---|---|---|
| G1 | | 16 ± 4 | | | | | 4 ± 1 | 20 |
| G2 | | | | 0.16 ± 0.010 | | | | 0.16 |
| G3 | | | | 0.25 ± 0.055 | | | | 0.25 |
| G4 | | 0.38 ± 0.060 | | 19 ± 3 | | | | 19 |
| G5 | | 0.12 ± 0.01 | | 0.56 ± 0.07 | | | | 0.69 |
| G6 | | 0.17 ± 0.011 | | 0.17 ± 0.02 | | | | 0.31 |
| G7 | | 32 ± 4 | | 0.13 ± 0.02 | | | | 33 |
| G8 | | | | 0.23 ± 0.028 | | | | 0.23 |
| G9 | | 32 ± 5 | | | 1.8 ± 0.3 | | | 34 |
| G10 | | | | | | | | 0 |
| G11 | | | | 0.10 ± 0.02 | | | | 0.10 |
| G12 | | 0.10 ± 0.01 | | | | | | 0.10 |
| G13 | | 0.08 ± 0.01 | | | | | 0.005 ± 0.001 | 0.087 |
| G14 | | 0.063 ± 0.001 | | | | | | 0.063 |
| G15 | | | | | | | | 0 |
| G16 | | | | | | | | 0 |
| G17 | | 0.077 ± 0.004 | | | | | | 0.077 |
| G18 | | 0.14 ± 0.01 | | | | | | 0.14 |
| G19 | | 1.26 ± 0.07 | | | | 0.036 ± 0.002 | | 1.30 |
| G20 | | 2.5 ± 0.2 | | | | | 0.33 ± 0.011 | 2.9 |
| G21 | | 0.035 ± 0.002 | | 0.128 ± 0.001 | | | | 0.16 |
| G22 | | | | | | | | 0 |
| G23 | | 0.048 ± 0.003 | | | | | | 0.048 |
| G24 | | 0.9 ± 0.1 | | 0.38 ± 0.03 | | | | 1.36 |
| G25 | | | | 0.053 ± 0.007 | | | | 0.0053 |
| G26 | | | | | | | | 0 |
| G27 | | | | | | | | 0 |
| G28 | | 0.37 ± 0.04 | | | | 0.08 ± 0.004 | | 0.40 |
| G29 | | 39 ± 1 | | 0.004 ± 0.001 | | 0.38 ± 0.01 | | 40 |
| G30 | | | | 0.011 ± 0.001 | | | | 0.0111 |
| G31 | | | | | | | | 0 |
| G32 | | | | | | | | 0 |
| G33 | | 56 ± 6 | 0.14 ± 0.01 | 0.31 ± 0.04 | | 0.15 ± 0.03 | | 57 |
| G34 | | 0.043 ± 0.005 | | 0.45 ± 0.05 | | | | 0.49 |
| G35 | | | | | | | | 0 |
| G36 | | 0.44 ± 0.02 | | 0.091 ± 0.006 | | | | 0.54 |
| G37 | | | | | | | | 0 |
| G38 | | 35.12 ± 0.02 | 0.010 ± 0.001 | | | 0.094 ± 0.001 | | 35.2 |
| G39 | | 0.032 ± 0.0039 | | | | | | 0.0332 |
| G40 | | | | 0.0370 ± 0.0005 | | | | 0.0370 |
| G41 | | | 0.017 ± 0.001 | | | | | 0.0172 |
| G42 | | 1.69 ± 0.018 | | | | | | 1.692 |
| G43 | | 0.09 ± 0.014 | | | | | | 0.0910 |
| G44 | | 0.10 ± 0.013 | | | | | | 0.102 |

Table 3. Cont.

| | Musk Xylene | Galaxolide | Phantolide | Cashmeran | Traseolide | Tonalide | Ambrettolide | Total Content |
|---|---|---|---|---|---|---|---|---|
| G45 | | | | | | | | 0 |
| G46 | | | | | | | | 0 |
| G47 | | | | | | | | 0 |
| G48 | | 0.14 ± 0.022 | | | | | | 0.140 |
| G49 | | | | | | | | 0 |
| G50 | | 0.241 ± 0.001 | | 0.17 ± 0.02 | | | | 0.415 |
| G51 | | 8.7 ± 0.8 | | 2.8 ± 0.2 | | | | 11.6 |
| G52 | | | | 0.013 ± 0.001 | | | | 0.0133 |
| G53 | | 0.017 ± 0.007 | 28.9 ± 0.6 | 0.8 ± 0.1 | | | | 29.8 |
| G54 | | 0.056 ± 0.001 | | | | | | 0.0567 |
| G55 | | 0.26 ± 0.04 | | | | | | 0.266 |
| G56 | | 0.18 ± 0.02 | | | | | | 0.188 |
| G57 | | 0.19 ± 0.098 | | 0.008 ± 0.001 | | | | 0.0202 |
| G58 | | 0.033 ± 0.001 | | 0.014 ± 0.005 | | | | 0.0471 |
| G59 | | 10 ± 3 | | | | | | 10.4 |
| G60 | | | | | | | | 0 |
| G61 | | 0.613 ± 0.006 | | | | | | 0.6133 |
| G62 | | | | | | | | 0 |
| G63 | | 0.31 ± 0.06 | | | | 0.0220 ± 0.0003 | | 0.332 |
| G64 | | 0.062 ± 0.003 | | | | | | 0.0621 |
| G65 | 2356 ± 31 | 0.05 ± 0.01 | | | | | | 2355 |
| G66 | 1840 ± 102 | 0.033 ± 0.001 | | | | | | 1840 |
| G67 | | 0.077 ± 0.001 | | 74.3 ± 0.2 | | | | 74.4 |

Only 20% of the samples were free of these compounds (16/67). The highest number of synthetic musks (galaxolide, cashmeran, phantolide and tonalide) were detected in sample G33. Although several of these PCPs such as galaxolide were found at concentrations above 10 µg g$^{-1}$ (e.g., 56 µg g$^{-1}$ in G33), in most cases, these compounds were presented at concentrations below 1 µg g$^{-1}$. An exception is the musk xylene which was detected in two samples at extraordinarily high concentrations, 2356 µg g$^{-1}$ in G65 and 1840 µg g$^{-1}$ in G66; this being the synthetic musk with the highest concentrations followed by cashmeran at 74 µg g$^{-1}$ in G67.

Regulatory Issues

The regulation of cosmetic products states that the synthetic musks ambrette, tibetene and moskene are prohibited. These chemicals were not detected in any of the analysed hydroalcoholic gels. Nevertheless, 51 samples contained at least one target synthetic musk, so the terms "parfum" or "aroma" must appear on their label. However, only 17 samples (see Table S2) were labelled with the word 'parfum', which indicates that 34 of 51 are under-labelled, although in general, the concentrations for this family of fragrances were quite low as it was commented. Some substances such as tonalide, phantolide and musk xylene must be mentioned when their concentration surpass 1%, 2% and 0.03%, respectively. Musk xylene was detected in two samples at very high concentrations, surpassing the regulation's limit, and its presence was not indicated in the label of any sample. In the case of phantolide and tonalide, they were detected in four and six samples, respectively, but none of them surpassed the regulation's limit, so they did not have to be indicated, as it is the case. As was mentioned, galaxolide was the most frequently found (63% of the analysed samples) and is under assessment as a persistent, bioaccumulative and endocrine disruptor [21].

### 3.3.3. Preservatives

All target preservatives were found in the analysed samples, as can be seen in Table 4.

**Table 4.** Individual concentration (μg g$^{-1}$) of the preservatives detected in 67 hydroalcoholic gel samples and total content.

| | PhEtOH | BHA | BHT | TCS | MeP | EtP | iPrP | PrP | iBuP | BzP | Total Content |
|---|---|---|---|---|---|---|---|---|---|---|---|
| G1 | | | 0.29 ± 0.01 | 0.3 ± 0.1 | | 50 ± 11 | | | 1.2 ± 0.34 | | 51.8 |
| G2 | | | 0.012 ± 0.001 | | | | | | | | 0.0121 |
| G3 | | | 0.009 ± 0.001 | | | | | | | | 0.0099 |
| G4 | | | 0.05 ± 0.01 | | 23 ± 4 | 29 ± 3 | | 3.92 ± 0.02 | 61 ± 1 | | 118 |
| G5 | | | | | 2.5 ± 0.27 | 7.2 ± 0.9 | | | | | 9.84 |
| G6 | | | | | | 1.0 ± 0.1 | | | | | 1.06 |
| G7 | | | | | | 1.8 ± 0.1 | | | | | 1.81 |
| G8 | 72.5 ± 0.5 | | | | | 2.3 ± 0.45 | | | | | 74.8 |
| G9 | | | | | 2.5 ± 0.8 | | | | | | 2.54 |
| G10 | | | | | 1.19 ± 0.01 | | | 150 ± 13 | | | 151 |
| G11 | | | | | | 1.0 ± 0.1 | | | | | 1.0 |
| G12 | | | | | | 0.6 ± 0.1 | | | | | 0.654 |
| G13 | 245 ± 2 | | 1.15 ± 0.01 | 0.04 ± 0.01 | | 2.6 ± 0.24 | | | | | 248 |
| G14 | | | | | | | | | | | 0 |
| G15 | 166 ± 15 | | | | | 0.58 ± 0.048 | | | | | 167 |
| G16 | | | | | | 0.64 ± 0.05 | | | 4.5 ± 0.2 | | 5.16 |
| G17 | | 0.105 ± 0.002 | | | | 62 ± 5 | | | 28 ± 3 | | 90.8 |
| G18 | | | | | | | | | 12.4 ± 0.8 | | 12.4 |
| G19 | | | | 0.10 ± 0.0010 | 2.57 ± 0.13 | 0.76 ± 0.03 | | | 23 ± 2 | | 26.4 |
| G20 | | | | | | | | | 3.6 ± 0.1 | | 3.62 |
| G21 | | | | 0.169 ± 0.003 | | 40 ± 11 | | | 1.7 ± 0.2 | | 42.3 |
| G22 | 10,477 ± 797 | | | | | | | | | | 10,477 |
| G23 | | | | | | | | 0.6 ± 0.2 | | | 0.644 |
| G24 | | | 170 ± 1 | 0.03 ± 0.01 | 1.05 ± 0.07 | 10 ± 1 | | | 48.81 ± 0.04 | 2.9 ± 0.30 | 233 |
| G25 | 0.5 ± 0.10 | | | | | | | | | | 0.551 |
| G26 | | | | | | | | | | | 0 |
| G27 | | | | 0.097 ± 0.005 | | 3.4 ± 0.7 | | | | | 3.53 |
| G28 | | | 0.034 ± 0.006 | | | | | | | | 0.0348 |
| G29 | | 0.12 ± 0.01 | 6.19 ± 0.04 | 0.088 ± 0.007 | 0.49 ± 0.015 | 4.24 ± 0.01 | | | 0.45 ± 0.01 | | 11.6 |
| G30 | | | | | 0.18 ± 0.01 | 1.5 ± 0.25 | | | | | 1.73 |
| G31 | 9352 ± 19 | | | | | | | | | | 9352 |
| G32 | | | | 0.136 ± 0.001 | | | | | | | 0.0136 |
| G33 | 1.217 ± 0.098 | | | | 5.4 ± 0.4 | | | 1.0 ± 0.19 | | | 7.71 |
| G34 | 65 ± 36 | | | | | 1.3 ± 0.2 | 0.3 ± 0.1 | | | | 66.6 |
| G35 | 7908 ± 815 | | | | | | | | | | 7908 |
| G36 | 16.3 ± 4.4 | | 2.902 ± 0.076 | | 1.142 ± 0.097 | 3.6 ± 0.2 | | | | | 24.0 |
| G37 | | | | | | 4 ± 1 | | | | | 3.8 |
| G38 | | 0.260 ± 0.002 | 0.0125 ± 0.0010 | | | 6.3 ± 0.4 | | | | | 6.57 |
| G39 | 3196 ± 871 | | | | | 0.8 ± 0.1 | | | | | 3197 |
| G40 | | | | | | | | | | | 0 |
| G41 | 0.55 ± 0.12 | | | | | | | | | | 0.551 |
| G42 | 7.0 ± 2.0 | | | | | 5.94 ± 0.04 | | | 0.243 ± 0.008 | | 13.2 |
| G43 | 0.212 ± 0.065 | | | | | 1.6 ± 0.3 | | | | | 1.85 |
| G44 | | | 0.056 ± 0.007 | | | 4.4 ± 0.5 | 0.13 ± 0.013 | | | | 4.60 |
| G45 | | | | | | 3.1 ± 0.4 | | | | | 3.18 |
| G46 | 0.6 ± 0.11 | | | | | | | | | | 0.61 |
| G47 | | | | | | | | | | | 0 |
| G48 | 1.6 ± 0.3 | | | | | 6.3 ± 0.87 | | | | | 7.96 |
| G49 | 13,735 ± 373 | | | 0.454 ± 0.001 | | | | 0.151 ± 0.006 | | | 13,735 |
| G50 | 19 ± 1 | 0.12 ± 0.02 | 0.003 ± 0.001 | | 1.9 ± 0.7 | 15 ± 2 | | | | | 36.4 |
| G51 | 21 ± 3 | 0.57 ± 0.04 | 0.020 ± 0.002 | 0.32 ± 0.09 | 22 ± 2 | | | 0.3 ± 0.11 | | | 44.5 |

Table 4. Cont.

| | PhEtOH | BHA | BHT | TCS | MeP | EtP | iPrP | PrP | iBuP | BzP | Total Content |
|---|---|---|---|---|---|---|---|---|---|---|---|
| G52 | 28 ± 7 | | | | | | | | | | 28.8 |
| G53 | 8 ± 1 | 0.35 ± 0.07 | 0.8 ± 0.1 | 0.05 ± 0.01 | 7.0 ± 0.92 | | | | 13.1 ± 0.6 | | 29.7 |
| G54 | | | | | | 9 ± 4 | | | | 17.3 ± 0.2 | 26.0 |
| G55 | 20.0 ± 6.5 | | | | 0.61 ± 0.07 | 5.4 ± 0.2 | | | | | 26.1 |
| G56 | 5428 ± 104 | | | | 1.21 ± 0.05 | 8 ± 1 | | | 59 ± 15 | | 5495 |
| G57 | | | | | | 3.4 ± 0.1 | | | | | 3.46 |
| G58 | | 0.008 ± 0.002 | | | | | | | | | 0.0086 |
| G59 | | | | | | 1.7 ± 0.5 | | | | | |
| G60 | | | | | | | | | | | 0 |
| G61 | | | 0.014 ± 0.001 | | | | | | | | 0.0145 |
| G62 | | | | | | 1.3 ± 0.10 | | | | 4.68 ± 0.14 | 6.02 |
| G63 | 1.00 ± 0.069 | 0.0124 ± 0.0003 | 0.008 ± 0.001 | | 0.50 ± 0.08 | | | | | | 1.54 |
| G64 | | | | | | | | | | | 0 |
| G65 | | | | 0.05 ± 0.017 | | | 0.13 ± 0.04 | | 0.7 ± 0.3 | | 0.942 |
| G66 | | | | 0.05 ± 0.02 | | | 0.070 ± 0.003 | | 0.14 ± 0.05 | | 0.302 |
| G67 | | | 4.3 ± 0.3 | | | 15 ± 3 | | | | | 19.3 |

The most frequently found were EtP in half of the samples (52%) and PhEtOH in 37% (25/67). The antioxidant BHT and the parabens MeP and iBuP were found in more than 20% of the samples and TCS and iBuP were detected in more than 10 samples.

The highest number of preservatives found in the same sample was six in G29, G51 and G53. Nevertheless, more than the 56% of the samples contained between 1 (25/67) and 2 (13/67) of the target preservatives. In six samples, none of the target preservatives were present (G14, G26, G40, G47, G60 and G64).

In most cases, the concentrations were at the low $\mu g\ g^{-1}$ (Table 4), excluding PhEtOH which was found at very high concentrations of up to 13,735 $\mu g\ g^{-1}$. The median concentration value of this substance in the analysed samples was 20 $\mu g\ g^{-1}$, which is the highest value with a noticeable difference with the rest of the preservatives. Regarding the detected parabens, PrP was presented at concentrations of up to 150 $\mu g\ g^{-1}$, EtP, iBuP and BzP up to 60 $\mu g\ g^{-1}$ and MeP up to 23 $\mu g\ g^{-1}$. BHT was detected at concentrations of up to 170 $\mu g\ g^{-1}$ and TCS and BHA only appear at very low concentrations below 1 $\mu g\ g^{-1}$.

Regulatory Issues

The Annex V of the EC Regulation No 1223/2009 comprised the preservatives allowed in cosmetic products, among which some of the target substances are included (PhEtOH, MeP, EtP and PrP).

The established maximum concentration in ready-for-use preparation for PhEtOH, a substance which is harmful if ingested, causing serious eye damage and may cause respiratory irritation, is 1.0% (10,000 $\mu g\ g^{-1}$). However, samples G49 and G22 surpassed this limit since they contained 1.4 and 1.0% ($w/w$) of the sample, respectively. The labels of these hydroalcoholic gels in addition to G35, G37 and G44 show that this compound is present at 2.1% ($w/w$). As described in Table S2, these samples comply with UNE-EN standards and, although all samples have the same labelling, the brand name is different, so they were analysed individually. The appearance of this compound in the product labelling in such high concentrations is due to these hydroalcoholic gels being considered biocides and that ECHA has endorsed the approval of PhEtOH as an active substance in type 1 biocidal product, which are those intended for human hygiene [5,22]. Nevertheless, they contained other substances such as fragrances, so they should be considered cosme-tics, although some of them would not comply with this regulation.

As regard the parabens EtP and MeP, PrP were detected at concentrations lower than the permitted limits, 0.4% (EtP and MeP) and 0.14% (PrP). The prohibited parabens in cosmetic products, as iPrP, iBuP and BzP, were found in around 9%, 21% and 6%, respectively, of the analysed hand sanitisers at concentrations of up to 60 µg g$^{-1}$ (iBuP in G4 and BzP in G56).

The antioxidant BHT, which has been very recently (July 2023) included in the Annex III of Regulation EC 1223/2009—substances in cosmetics products must not contain except subject to the restrictions laid down—is regulated for leave-on and rinse-off products at a maximum concentration of 8000 µg g$^{-1}$, but none of the hydroalcoholic gels surpassed this value since the highest concentration was 170 µg g$^{-1}$ in G24. In contrast, the triclosan is regulated but it does not have a limit for this type of cosmetic products.

### 3.3.4. Plasticisers

Considering plasticisers, four phthalates and three adipates were present in the analysed hydroalcoholic gels: DMP (8/67), DEP (24/67), DPP (3/67), DEHP (15/67) and DMA (7/67), DEA (10/67), DEHA (2/67), as can be seen in Table 5.

**Table 5.** Individual concentration (µg g$^{-1}$) of plasticisers detected in 67 hydroalcoholic gel samples and total content.

| | DMP | DEP | DPP | DEHP | DMA | DEA | DEHA | Total Content |
|---|---|---|---|---|---|---|---|---|
| G1 | | | | 0.4 ± 0.13 | 0.22 ± 0.01 | 1.0 ± 0.3 | | 1.73 |
| G2 | | 1.70 ± 0.07 | | | 0.10 ± 0.01 | | | 1.82 |
| G3 | | 0.230 ± 0.0034 | | 0.12 ± 0.03 | 0.22 ± 0.04 | | | 0.586 |
| G4 | | | | 0.15 ± 0.02 | | | | 0.155 |
| G5 | 0.42 ± 0.05 | | | 0.114 ± 0.004 | 4.0 ± 0.1 | | | 4.61 |
| G6 | 0.59 ± 0.010 | 25.5 ± 0.92 | | | | | | 26.1 |
| G7 | | 3.2 ± 0.2 | | | | | | 3.25 |
| G8 | | 0.19 ± 0.02 | | | | | | 0.191 |
| G9 | 10 ± 2 | 104 ± 7 | | | | | | 113.8 |
| G10 | | 1.8 ± 0.2 | | | | | | 1.87 |
| G11 | | | | | | | | 0 |
| G12 | | | | | | | | 0 |
| G13 | 0.45 ± 0.010 | 1804 ± 123 | | | | 0.15 ± 0.013 | | 1804 |
| G14 | | | | | | | | 0 |
| G15 | | | | | | | | 0 |
| G16 | | | | | | | | 0 |
| G17 | | 8.7 ± 1.2 | | | | | | 8.7 |
| G18 | | | | | | | | 0 |
| G19 | 0.53 ± 0.07 | | | 1.1 ± 0.1 | | 1.37 ± 0.02 | | 3.09 |
| G20 | | 8238 ± 783 | | | | | 0.399 ± 0.007 | 8238 |
| G21 | | | | | | | | 0 |
| G22 | | | | | | | | 0 |
| G23 | | 0.68 ± 0.30 | | | | | | 0.68 |
| G24 | | | | | | | | 0 |
| G25 | | | | | | | | 0 |
| G26 | | | | | | | | 0 |
| G27 | | | | | | | | 0 |
| G28 | 108 ± 4 | | | | | 0.098 ± 0.01 | | 108 |
| G29 | | 6.3 ± 0.3 | | | | 0.11 ± 0.01 | | 6.51 |
| G30 | | | | | | 0.047 ± 0.0012 | | 0.047 |
| G31 | 0.63 ± 0.09 | 4888 ± 13 | | | | | | 4888 |

Table 5. Cont.

|  | DMP | DEP | DPP | DEHP | DMA | DEA | DEHA | Total Content |
|---|---|---|---|---|---|---|---|---|
| G32 |  | 6.5 ± 0.7 |  |  |  |  |  | 6.5 |
| G33 |  |  |  |  |  |  |  | 0 |
| G34 |  | 0.19 ± 0.05 |  |  | 0.7 ± 0.1 |  |  | 0.85 |
| G35 |  |  |  |  |  |  |  | 0 |
| G36 |  | 0.24 ± 0.03 |  | 0.14 ± 0.036 | 0.56 ± 0.027 |  |  | 0.96 |
| G37 |  |  |  |  |  |  |  | 0 |
| G38 |  | 8 ± 1 |  | 0.25 ± 0.045 |  |  |  | 8.58 |
| G39 |  | 0.2 ± 0.1 |  | 0.14 ± 0.04 | 0.212 ± 0.005 | 0.010 ± 0.007 |  | 0.569 |
| G40 |  | 0.231 ± 0.007 |  |  |  |  |  | 0.231 |
| G41 |  |  |  |  |  |  |  | 0 |
| G42 |  |  |  |  |  |  |  | 0 |
| G43 |  |  |  |  |  |  |  | 0 |
| G44 |  |  |  | 0.9 ± 0.2 |  | 0.053 ± 0.001 |  | 0.957 |
| G45 |  |  |  | 2.8 ± 0.2 |  |  |  | 2.8 |
| G46 |  |  |  |  |  |  |  | 0 |
| G47 |  |  |  |  |  |  |  | 0 |
| G48 |  |  |  |  |  |  |  | 0 |
| G49 |  |  |  |  |  |  |  | 0 |
| G50 |  | 0.79 ± 0.05 |  |  |  |  |  | 0.79 |
| G51 | 0.16 ± 0.01 | 4 ± 1 |  | 2.1 ± 0.3 |  | 0.080 ± 0.001 | 0.9 ± 0.2 | 7.79 |
| G52 |  |  |  |  |  |  |  | 0 |
| G53 |  | 1910 ± 106 |  | 12 ± 3 |  |  |  | 1922 |
| G54 |  |  |  |  |  |  |  | 0 |
| G55 |  |  |  |  |  |  |  | 0 |
| G56 |  | 0.6 ± 0.1 | 0.11 ± 0.02 | 0.19 ± 0.05 |  |  |  | 0.952 |
| G57 |  |  |  |  |  |  |  | 0 |
| G58 |  |  |  |  |  |  |  | 0 |
| G59 |  |  |  |  |  |  |  | 0 |
| G60 |  |  |  |  |  |  |  | 0 |
| G61 |  |  |  | 0.24 ± 0.02 |  |  |  | 0.24 |
| G62 |  | 538 ± 4 |  |  |  |  |  | 538 |
| G63 |  |  |  | 0.333 ± 0.094 |  |  |  | 0.333 |
| G64 |  |  |  |  |  |  |  | 0 |
| G65 |  |  | 0.05 ± 0.01 |  |  |  |  | 0.050 |
| G66 |  |  | 0.14 ± 0.06 |  |  | 1.77 ± 0.08 |  | 1.92 |
| G67 |  |  |  |  |  |  |  | 0 |

Regarding the number of compounds, in most samples, two or three of the target compounds were found. One sample presented five of the compounds (G51) followed by the sample G39 with four compounds (Table 5). Almost half of the samples (31) were free of the target plasticisers.

In general, the concentrations were below 2 µg g$^{-1}$ but it should be underlined that there is the high concentration of DEP in some samples with concentrations between 1800 and 8240 µg g$^{-1}$ (0.2–1%). DMP reached a concentration of 108 µg g$^{-1}$ in sample G28, although this value is much lower than the previous values indicated for DEP.

Regulatory Issues

Some of the analysed plasticisers are banned by the EC Regulation No 1223/2009. Among the detected ones, DEHP and DPP are prohibited in cosmetics (see Figure 3), and they were detected in 3 and 15 samples, respectively. Both substances are toxic for reproduction and DEHP is also considered an endocrine disruptor. The presence of these phthalates may be due to migration from the plastic container to the hydroalcoholic gel. Previous studies have demonstrated the transfer of certain substances from the applicators to cosmetic products [23]. Other prohibited substances such as DBP, DMEP, DIPP, BBP, DIBP, DIHP, DPhP, DnOP and DCHP were not found in any of the analysed samples. On the other hand, it is important to note that DEP, the plasticiser found in the highest concentrations (up to 8238 µg g$^{-1}$), is under evaluation as an endocrine disruptor, even though it is not legislated in cosmetic products.

## 4. Conclusions

Sixty-seven hydroalcoholic gels collected in different establishments were analysed by SPME-GC-MS/MS, a validated methodology which allows a complete analysis of the hydroalcoholic gel samples. Sixty personal care products, including fragrance allergens, synthetic musks, preservatives and plasticisers were targeted. The results revealed the presence of 48 out of these 60 target compounds in the analysed hydroalcoholic gels. Only one sample complied with the WHO recommendations for hand sanitiser formulations. The highest concentrations were observed for fragrance allergens and many of the samples did not comply with the legislation for these substances as, despite not being labelled, they contained some above the 0.001% limit in leave-on products. Furthermore, some prohibited compounds (iPrP, iBuP, BzP, DEHP and DPP) were detected in some cases. In addition, some hydroalcoholic gels are considered biocidal, although after the analysis of their composition, they should be considered cosmetic products. In this context, PhEtOH sometimes exceeds the limits set for cosmetics, while it is allowed for biocides for human hygienic use. Therefore, results demonstrated that greater control over the formulations of these frequently used cosmetic products is necessary to ensure consumer safety without causing undesirable side effects.

**Supplementary Materials:** The following supporting information can be downloaded at: https://www.mdpi.com/article/10.3390/mps6050095/s1, Table S1: Target compounds. CAS number, molecular mass, retention time, MS/MS transitions and restrictions for leave-on cosmetics (EC 1223/2009 regulation) [6]; Table S2: Information for the analysed samples including the composition indicated on the label; Figure S1: Comparison of the recoveries obtained for fortified hydroalcoholic gel samples free of target compounds (except DEP) at two levels: 0.2 µg g$^{-1}$ and 2 µg g$^{-1}$. Table S3: SPME-GC-MS/MS performance. Linearity, precision, recoveries, instrumental detection limits (IDLs) and limits of detection (LODs).

**Author Contributions:** Conceptualisation, M.L.; validation, A.C.-L. and L.V.; formal analysis, A.C.-L. and L.V.; investigation, A.C.-L. and L.V.; resources, M.L. and T.D.; data curation, A.C.-L. and L.V.; writing—original draft preparation, A.C.-L. and M.C.; writing—review and editing, M.C. and M.L.; supervision, M.L.; project administration, M.L.; funding acquisition, M.L. and T.D. All authors have read and agreed to the published version of the manuscript.

**Funding:** This research was funded by the projects ED431 2020/06, IN607B 2022/15 (Consolidated Research Groups Program, Xunta de Galicia). This study was based upon work from the Sample Preparation Study Group and Network, supported by the Division of Analytical Chemistry of the European Chemical Society. The authors are members of the National Network for Sustainability in Sample Preparation, RED2022-134079-T (Ministry of Science, Innovation and Universities, Spain). These programmes are co-funded by FEDER (EU).

**Institutional Review Board Statement:** Not applicable.

**Informed Consent Statement:** Not applicable.

**Data Availability Statement:** All data are available in the manuscript and Supplementary Material.

**Acknowledgments:** A.C.-L. acknowledges the Xunta de Galicia predoctoral contract (ED481A and IN606A). M.C. and M.L. acknowledge the IUPAC project 2021-015-2-500: Greenness of official standard sample preparation methods.

**Conflicts of Interest:** The authors declare no conflict of interest.

## References

1. Lotfinejad, N.; Peters, A.; Pittet, D. Hand hygiene and the novel coronavirus pandemic: The role of healthcare workers. *J. Hosp. Infect.* **2020**, *105*, 776. [CrossRef] [PubMed]
2. Berardi, A.; Perinelli, D.R.; Merchant, H.A.; Bisharat, L.; Basheti, I.A.; Bonacucina, G.; Cespi, M.; Palmieri, G.F. Hand sanitisers amid COVID-19: A critical review of alcohol-based products on the market and formulation approaches to respond to increasing demand. *Int. J. Pharm.* **2020**, *584*, 119431. [CrossRef] [PubMed]
3. Huddleston, T., Jr. The History of Hand Sanitize-How the Coronavirus Staple Went from Mechanic Shops to Consumer Shelves. Available online: https://www.cnbc.com/2020/03/27/coronavirus-the-history-of-hand-sanitizer-and-why-its-im-portant.html (accessed on 25 January 2023).
4. European Commission. Guidance on the Applicable Legislation for Leave-on Hand Cleaners and Hand Disinfectants (Gel, Solution, etc.). Available online: https://ec.europa.eu/docsroom/documents/40523 (accessed on 5 June 2023).
5. European Commission. Regulation (EU) No 528/2012 of the European Parliament and of the Council of 22 May 2012 concerning the making available on the market and use of biocidal products. *Off. J. Eur. Union* **2012**, *L167*, 1–123.
6. European Commission. Regulation (EC) No 1223/2009 of the European Parliamen and of the Council of 30 November 2009 on cosmetic products. *Eur. Off. J.* **2009**, *L342*, 59–209.
7. World Health Organization. Guide to Local Production: WHO-Recommended Handrub Formulations. 2010. Available online: https://www.who.int/publications/i/item/WHO-IER-PSP-2010.5 (accessed on 8 May 2023).
8. Reeder, M.J. Allergic contact dermatitis to fragrances. *Dermatol. Clin.* **2020**, *38*, 371–377. [CrossRef] [PubMed]
9. Nicolopoulou-Stamati, P.; Hens, L.; Sasco, A.J. Cosmetics as endocrine disruptors: Are they a health risk? *Rev. End. Metab. Dis.* **2015**, *16*, 373–383. [CrossRef] [PubMed]
10. Yazar, K.; Johnsson, S.; Lind, M.L.; Boman, A.; Liden, C. Preservatives and fragrances in selected consumer-available cosmetics and detergents. *Contact Dermat.* **2011**, *64*, 265–272. [CrossRef] [PubMed]
11. Al-Mussallam, A.S.; Bawazir, A.T.; Alshathri, R.S.; Alharthi, O.; Aldawsari, F.S. Optimization of a gas chromatography-mass spectrometry (GCMS) method for detecting 28 allergens in various personal care products. *Cosmetics* **2023**, *10*, 91. [CrossRef]
12. Fallica, F.; Leonardi, C.; Toscano, V.; Santonocito, D.; Leonardi, P.; Puglia, C. Assessment of Alcohol-based hand sanitizers for long-term use, formulated with addition of natural ingredients in comparison to WHO formulation 1. *Pharmaceutics* **2021**, *13*, 571. [CrossRef] [PubMed]
13. Chojnacki, M.; Dobrotka, C.; Osborn, R.; Johnson, W.; Young, M.; Meyer, B.; Laskey, E.; Wozniak, R.A.F.; Dewhurst, S.; Dunman, P.M. Evaluating the antimicrobial properties of commercial hand sanitizers. *MSphere* **2021**, *6*, 21. [CrossRef] [PubMed]
14. Kratzel, A.; Todt, D.; V'kovski, P.; Steiner, S.; Gultom, M.; Thao, T.T.N.; Ebert, N.; Holwerda, M.; Steinmann, J.R.; Niemeyer, D. Inactivation of severe acute respiratory syndrome coronavirus 2 by WHO-recommended hand rub formulations and alcohols. *Emerg. Infe. Dis.* **2020**, *26*, 1592. [CrossRef] [PubMed]
15. Alvarez-Rivera, G.; Vila, M.; Lores, M.; Garcia-Jares, C.; Llompart, M. Development of a multi-preservative method based on solid-phase microextraction-gas chromatography-tandem mass spectrometry for cosmetic analysis. *J. Chromatogr. A* **2014**, *1339*, 13–25. [CrossRef] [PubMed]
16. Celeiro, M.; Garcia-Jares, C.; Llompart, M.; Lores, M. Recent advances in sample preparation for cosmetics and personal care products analysis. *Molecules* **2021**, *26*, 4900. [CrossRef] [PubMed]
17. Vazquez, L.; Celeiro, M.; Castiñeira-Landeira, A.; Dagnac, T.; Llompart, M. Development of a solid phase microextraction gas chromatography tandem mass spectrometry methodology for the analysis of sixty personal care products in hydroalcoholic gels—hand sanitizers—in the context of COVID-19 pandemic. *Anal. Chim. Acta* **2022**, *1203*, 339650. [CrossRef] [PubMed]
18. Kavian, M.; Ghani, M.; Raoof, J.B. Polyoxometalate/reduced graphene oxide composite stabilized on the inner wall of a stainless steel tube as a sorbent for solid-phase microextraction of some parabens followed by quantification via high-performance liquid chromatography. *Microchem. J.* **2023**, *187*, 108413. [CrossRef]
19. Wojnowski, W.; Tobiszewski, M.; Pena-Pereira, F.; Psillakis, E. AGREEprep-Analytical greenness metric for sample preparation. *TrAC-Trend. Anal. Chem.* **2022**, *149*, 116553. [CrossRef]
20. ECHA Lilial. Available online: https://echa.europa.eu/es/substance-information/-/substanceinfo/100.001.173 (accessed on 2 February 2023).
21. ECHA Galaxolide. Available online: https://echa.europa.eu/es/substance-information/-/substanceinfo/100.013.588 (accessed on 2 February 2023).

22. ECHA 2-Phenoxyethanol. Available online: https://echa.europa.eu/es/substance-information/-/substanceinfo/100.004.173 (accessed on 2 February 2023).
23. Lores, M.; Celeiro, M.; Rubio, L.; Llompart, M.; Garcia-Jares, C. Extreme cosmetics and borderline products: An analytical-based survey of European regulation compliance. *Anal. Bioanal. Chem.* **2018**, *410*, 7085–7102. [CrossRef] [PubMed]

**Disclaimer/Publisher's Note:** The statements, opinions and data contained in all publications are solely those of the individual author(s) and contributor(s) and not of MDPI and/or the editor(s). MDPI and/or the editor(s) disclaim responsibility for any injury to people or property resulting from any ideas, methods, instructions or products referred to in the content.

*Article*

# An Efficient, Simultaneous Electrochemical Assay of Rosuvastatin and Ezetimibe from Human Urine and Serum Samples

Leyla Karadurmus [1], Sevinc Kurbanoglu [2], Bengi Uslu [2] and Sibel A. Ozkan [2,*]

1. Department of Analytical Chemistry, Faculty of Pharmacy, Adıyaman University, Adıyaman 02040, Turkey
2. Department of Analytical Chemistry, Faculty of Pharmacy, Ankara University, Ankara 06560, Turkey
* Correspondence: ozkan@pharmacy.ankara.edu.tr

**Abstract:** The drug combination of rosuvastatin (ROS) and ezetimibe (EZE) is used to treat hypercholesterolemia. In this work, a simultaneous electrochemical examination of ROS and EZE was conducted for the first time. The electrochemical determination of ROS and EZE was carried out using adsorptive stripping differential pulse voltammetry (AdSDPV) on a glassy carbon electrode (GCE) in 0.1 M $H_2SO_4$. The effects of the pH, scan rate, deposition potential, and time on the detection of ROS and EZE were analyzed. Under optimum conditions, the developed sensor exhibited a linear response between $1.0 \times 10^{-6}$ M and $2.5 \times 10^{-5}$ M for EZE and $5.0 \times 10^{-6}$ M, and $1.25 \times 10^{-5}$ M for ROS. The detection limits for ROS and EZE were $3.0 \times 10^{-7}$ M and $2.0 \times 10^{-6}$ M, respectively. The developed sensor was validated in terms of linear range, accuracy, precision, the limit of determination (LOD), and the limit of quantification (LOQ), and it was evaluated according to ICH Guidelines and USP criteria. The proposed method was also used to determine ROS and EZE in human urine and serum samples, which are reported in terms of recovery studies.

**Keywords:** rosuvastatin; ezetimibe; glassy carbon electrode; adsorptive stripping differential pulse voltammetry

**Citation:** Karadurmus, L.; Kurbanoglu, S.; Uslu, B.; Ozkan, S.A. An Efficient, Simultaneous Electrochemical Assay of Rosuvastatin and Ezetimibe from Human Urine and Serum Samples. *Methods Protoc.* **2022**, *5*, 90. https://doi.org/10.3390/mps5060090

Academic Editors: Victoria Samanidou, Verónica Pino and Natasa Kalogiouri

Received: 29 September 2022
Accepted: 27 October 2022
Published: 1 November 2022

**Publisher's Note:** MDPI stays neutral with regard to jurisdictional claims in published maps and institutional affiliations.

**Copyright:** © 2022 by the authors. Licensee MDPI, Basel, Switzerland. This article is an open access article distributed under the terms and conditions of the Creative Commons Attribution (CC BY) license (https://creativecommons.org/licenses/by/4.0/).

## 1. Introduction

Rosuvastatin (ROS) is a hypercholesterolemia drug that lowers plasma cholesterol levels (Scheme 1a) [1]. ROS has a structure that is similar to most other synthetic statins, but unlike other statins, it contains sulfur. ROS is a competitive inhibitor of the enzyme HMG-CoA reductase [2–4]. Ezetimibe (EZE) is a drug that the FDA has confirmed as curing hypercholesterolemia (Scheme 1b). EZE is the first lipid-lowering drug that reduces the amount of lipoprotein cholesterol by preventing the absorption of cholesterol at the brush-border level of the intestine. It prevents the intestinal uptake of dietary and bile cholesterol [4,5].

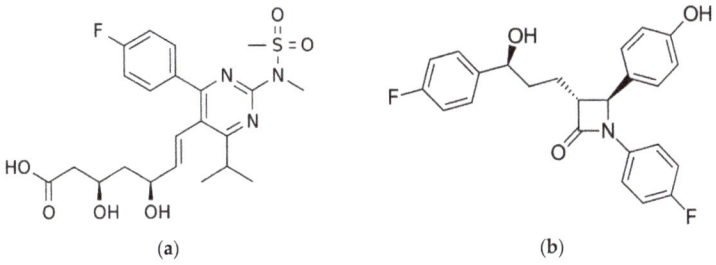

**Scheme 1.** The chemical structures of (**a**) rosuvastatin and (**b**) ezetimibe.

Statins and EZE have different lipid-lowering mechanisms of action, and combining them can obtain the strongest impact on lowering lipids and stabilizing plaque areas [6]. In the literature, it has been found that the combination of ROS and EZE further lowers total cholesterol and LDL cholesterol, clearly lowering triglyceride levels, and potentiating the lipid-lowering effects. The combination of ROS and EZE decreases lipid levels and the plaque burden. The combination of a statin and EZE has a greater effect on coronary plaque regression in patients with acute coronary syndrome [6,7]. Adding EZE to ROS significantly improves many more lipid parameters than does doubling the ROS dose [8]. The literature includes descriptions of patients who received 5, 10, 20, or 40 mg of ROS every day, and the average plasma concentration for ROS was 1.6 ng/mL, 3.5 ng/mL, 6.3 ng/mL, and 9.8 ng/mL, respectively [9]. For patients taking one dose of 10 mg of ezetimibe, average ezetimibe peak plasma concentrations ($C_{max}$) of 3.4 to 5.5 ng/mL were acquired within 4 to 12 h [10].

The two major fields of the natural sciences, chemistry and electrical science, came together in the 19th century to form electrochemistry [11]. Electrochemical techniques are extensively used in drug analysis. Among all of the electrochemical methods, stripping analysis is one of the most sensitive electrochemical techniques, and it is therefore used in quantitative determinations, especially in drug analysis. In recent years, stripping voltammetry has been used in the analysis of many drug substances [12]. The reason for this great sensitivity is the combination of an efficient accumulation phase with advanced measurement processes that produce an excellent signal [13,14]. The adsorptive accumulation is intended to deposit the analyte present in the solution on an electrode surface with a small surface area. Stripping voltammetry is also used in clinical practice and allows the conduct of various analyses of human blood, urine, and tissues [15].

The literature reveals some analytical techniques for the simultaneous detection of ROS and EZE. These methods are reverse-phase high-performance liquid chromatography [16,17], micellar liquid chromatography [18], high-performance column liquid chromatography, high-performance thin-layer chromatography [19], spectrophotometry [20,21], and liquid chromatography/mass spectrometry [22,23]. In this work, EZE and ROS were electrochemically analyzed using the AdSDPV technique at GCE. The efficacy of the electrochemical method was fully analyzed for the detection of ROS and EZE in commercial human serum and in urine samples, and we report on it in terms of recovery studies.

## 2. Experimental Design

### 2.1. Materials

Different supporting electrolytes of $H_2SO_4$ solutions (0.1 and 0.5 M), acetate (pH 3.7–5.7), and phosphate (pH 2.0–8.0) buffers were prepared for electrochemical measurements. AdSDPV voltammogram recordings were obtained after the addition of each aliquot. Drug-free human serum from male AB plasma was purchased from Sigma-Aldrich (St. Louis, MO, USA). Acetic acid, acetonitrile, methanol, phosphoric acid, sodium acetate trihydrate, sodium dihydrogen phosphate dihydrate, sodium hydroxide, sodium phosphate monobasic, sodium phosphate, and sulfuric acid were purchased from Sigma-Aldrich. All reagents were of analytical grade and were used without pre-processing. All measurements were realized at room temperature; all solutions were kept from light and used within 24 h to prevent degradation.

### 2.2. Equipment

A Bioanalytical Systems (BAS 100W) electrochemical analyzer with a standard three-electrode system was used for the voltammetric measurements. The three-electrode system included a platinum-wire counter electrode, an Ag/AgCl-saturated KCl reference electrode, and a GCE (GC, BAS; 3 mm, diameter), which served as a working electrode. The surface of the GCE was polished with an aqueous slurry of alumina powder (Φ: 0.01 μm) on a damp, smooth polishing cloth just before each experiment. The pH was checked using a pH meter Model 538 (Weilheim, Germany). Operating conditions for AdSDPV were as follow: pulse amplitude, 50 mV; deposition time, 15 s; scan rate, 20 mV/s; pulse width, 50 ms; sensitivity, 10 µA/V; sample width, 17 ms; pulse period, 200 ms; quiet time, 10 s.

## 3. Procedures

### 3.1. Standards and Sample Preparation

The $1 \times 10^{-3}$ M stock solution of ROS and EZE was prepared in methanol and kept in a refrigerator (+4 °C). The solutions of ROS and EZE for the voltammetric measurements were prepared by direct dilution of the stock solution with 0.1 M $H_2SO_4$, and they included a constant amount of methanol (20%, *v:v*). Analytical curves were obtained by adding aliquots of the stock solutions of ROS and EZE into the electrochemical cell containing 10.0 mL of the 0.1 M $H_2SO_4$ with a constant amount of methanol.

### 3.2. Biological Sample Preparation

The applicability of the developed procedure to human urine samples was also investigated. Drug-free urine samples were collected from a healthy laboratory employee on the day of the experiment. To prepare a stock urine solution, 5.4 mL of acetonitrile, 3.6 mL of the drug-free urine samples, and 1 mL of the ROS/EZE stock solution ($1 \times 10^{-3}$ M) were placed in a 10 mL centrifuge tube. First, the mixture was vortexed for 10 min, and then it was centrifuged at 3500 rpm for 30 min. The supernatant part was carefully transferred to a distinct, clean tube. In this procedure, acetonitrile acted as a precipitating agent. A ROS/EZE-free sample of the same urine was used as a blank solution. All measurements were performed at least in triplicate, and the standard addition technique was performed for the determination of ROS/EZE.

Synthetic human serum was kept frozen at −20 °C in a freezer until analysis. For the preparation of a stock serum sample, a standard procedure was followed. Quantities of 1 mL of ROS/EZE, 5.4 mL of acetonitrile, and 3.6 mL of synthetic human serum were added to a centrifuge tube to prepare a stock serum solution. First, it was vortexed for 10 min and then centrifuged at 3500 rpm for 30 min, and later, the supernatant was taken. Here, acetonitrile was used to precipitate serum proteins. The supernatant was diluted with 0.1 M $H_2SO_4$ to prepare certain concentrations for the recovery measurements. All of the experiments were performed at least three times for calibration and five times for the recovery experiments.

Analytical curves were obtained by adding aliquots of the stock solutions of ROS and EZE from synthetic human serum or human urine into the electrochemical cell containing 10.0 mL of the 0.1 M $H_2SO_4$ with a constant amount of methanol.

## 4. Results and Discussion

### 4.1. Voltammetric Behavior of ROS and EZE

The voltammetric behavior of ROS and EZE was examined on a GCE in detail. In the first step, the behavior of ROS and EZE was investigated by CV studies to characterize their electrochemical oxidation behavior in the range of 0 V to 1.6 V. The CV results indicated the irreversible nature of the oxidation process of ROS and EZE. Moreover, the adsorptive stripping differential pulse voltammetric (AdSDPV) technique was further used, and the anodic oxidation was observed until reaching a potential of about 0.9 V, and 1.2 V; there was a single well-defined and sharp oxidation peak for EZE and ROS, respectively, using the AdSDPV technique on a GCE in 0.1 M $H_2SO_4$ (Figure 1).

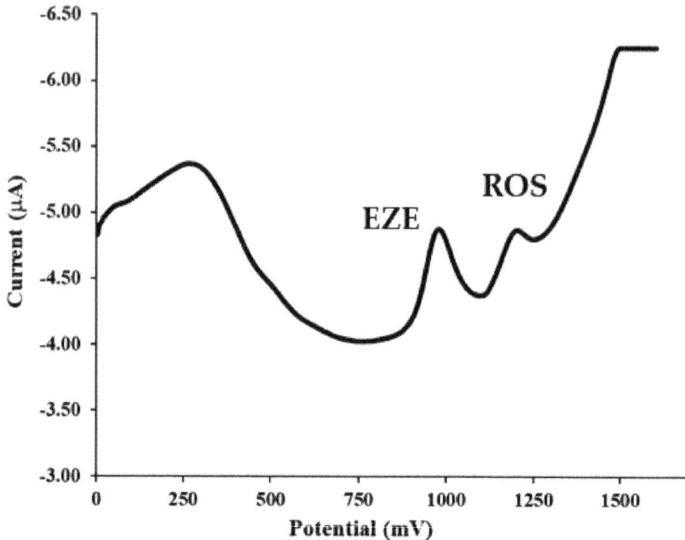

**Figure 1.** The AdSDPV voltammograms of $5.0 \times 10^{-6}$ M EZE and $7.5 \times 10^{-6}$ M ROS in 0.1 M $H_2SO_4$ (stripping conditions: accumulation potential of 0.0 V and accumulation time of 15 s).

### 4.2. Influence of the pH

The electrochemical behavior of ROS and EZE was studied within a wide pH range (pH 0.3–7.0) using the DPV technique on a GCE. With the DPV method, the maximum current occurred in the 0.1 M $H_2SO_4$ medium. The following equation followed the effect of pH on the peak potential. The $E_p$-pH plots indicated that a pH increase caused the shifting of peak potentials to less positive values (Figure 2).

$$E_p \text{ (mV)} = 1354.24 - 22.79 \text{ pH}; R^2 = 0.997 \text{ for ROS}$$

$$E_p \text{ (mV)} = 998.49 - 50.99 \text{ pH}; R^2 = 0.998 \text{ for EZE}$$

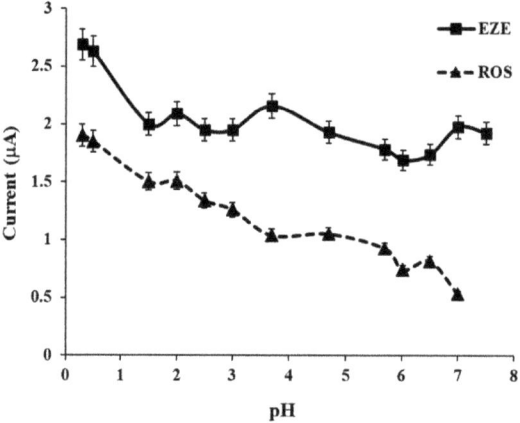

**Figure 2.** Plot of $I_p$ vs. pH of $1 \times 10^{-4}$ M ROS and EZE solution using the DPV.

## 4.3. Influence of the Scan Rate

Scan rate experiments were performed to understand the electrochemical oxidation/reduction mechanisms, such as adsorption or diffusion. The influence of the scan rate between 5 and 1000 mV/s on the peak current and potential was investigated in 0.1 M $H_2SO_4$ using CV, where the highest peak was obtained in pH studies using a GC electrode.

The plot of $E_p$ vs. log $v$ was linear; this attitude is coherent with the EC nature of the reaction in which the electrode reaction is coupled with an irreversible follow-up chemical step in CV. According to [24], $E_p$ can be defined by the following equation;

$$E_p = E^{0'} - \frac{2.303RT}{\alpha nF} \log \frac{RTk^0}{\alpha nF} + \frac{2.303RT}{\alpha nF} \log v$$

where $E^0$ is the formal potential, $R$ is the gas constant, $T$ is the temperature, $k^0$ is the standard heterogeneous rate constant, $\alpha$ is the transfer coefficient of the oxidation of ROS and EZE, $v$ is scan rate, $F$ is the Faraday constant, and n is the number of electrons that are involved in the electrooxidation of ROS and EZE [22].

In general, $\alpha$ is used as 0.5 for irreversible processes. Since $\alpha$ is 0.5 for irreversible systems, n can be calculated from

$E_p$ (V) = 0.046 log $v$ (V·s$^{-1}$) + 1.312 (r = 0.997) (0.1 × 10$^{-3}$ M ROS), and n is found to be 2.36 for ROS, and

$E_p$ (V) = 0.049 log $v$ (V·s$^{-1}$) + 1.066 (r = 0.997) (0.1 × 10$^{-3}$ M EZE), and n was calculated as being 2.38 for EZE (Figure 3a,b).

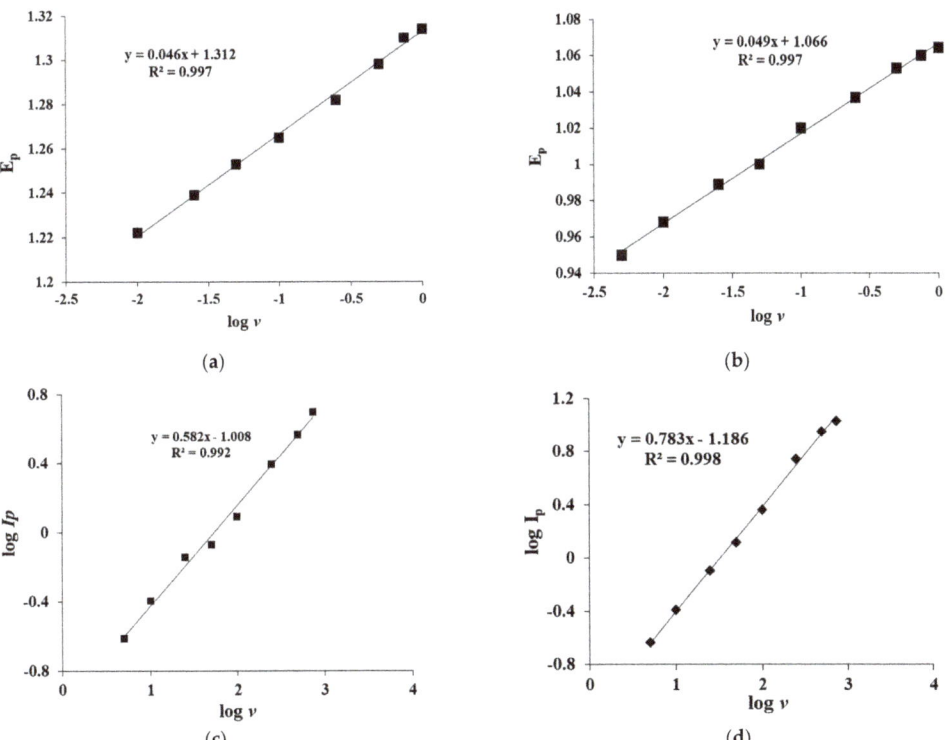

**Figure 3.** Results of scan rate studies: (**a**) $E_p$ vs. log $v$, (**b**) log $I_p$ vs. log $v$ for 0.1 mM ROS, (**c**) $E_p$ vs. log $v$, and (**d**) log $I_p$ vs. log $v$ for 0.1 mM EZE in $H_2SO_4$.

Moreover, the logarithm of peak current vs. the logarithm of scan rate gives more detailed information about the electrochemical mechanisms. When these graphs were plotted, for EZE, from the slope of the equation log $(I_p)$ = 0.783 log $v$ − 1.186 (r = 0.998), it can be concluded that the reaction is adsorption-controlled since the slope was close to 1. Thus, as a result of the scan rate experiments, in the 0.1 M $H_2SO_4$ medium, the electrochemical behavior of EZE was found to be adsorption-controlled (Figure 3d).

For ROS, the slope of the equation log $(I_p)$ = 0.582 log $v$ − 1.008 (r = 0.992), and the electrochemical behavior of ROS was found to be diffusion-controlled (Figure 3c). As we aimed to determine these two drug-active compounds simultaneously, we applied the adsorptive stripping method, which enabled us to assess ROS and EZE precisely.

In the literature, the electrochemical determination of ROS and its possible oxidation mechanism have been studied. The authors suggested an electrooxidation mechanism involving a Kolbe electrolysis reaction of the carboxylic acid group localized at the dihydroxyhept-6-enoic acid portion of the rosuvastatin calcium molecule [25–27]. In the literature, the electrochemical behavior and possible oxidation mechanism of EZE was also reported by the authors as being due to the inductive effect of the fluoride group in the aromatic rings of the EZE molecule; oxidation takes place in the hydroxyl group of phenol (EC mechanism) and the main voltammetric behavior of aromatic hydroxyl derivatives, which are structurally related to the mechanism of oxidation of EZE, may be postulated by the oxidation of the hydroxyl group on the aromatic ring [28,29].

### 4.4. Effect of Deposition Time and Potential

Parameters, such as deposition time and potential, significantly affect the AdSDPV peaks of the analytes. Hence, these parameters as related to AdSDPV were optimized to obtain the best results for the determination of ROS and EZE. The effect of the deposition time on stripping peak current was studied in the range of 0 s to 50 s, with 0 V deposition potential. It was observed that the peak current increased between 0 and 15 s (Figure 4). However, after 15 s, a decrease was observed in the peak current. As a result, 15 s was selected as the optimum deposition time. The effect of deposition potential, which is another important parameter, on stripping peak currents was studied in deposition potentials ranging from −0.1 V to +0.1 V, with a constant accumulation time of 15 s (Figure 4). A decrease in stripping peak currents was observed after the 0 V deposition potential, with an accumulation time of 15 s. A deposition time of 15 s and a deposition potential of 0 V, at which the maximum peak current was observed, were used in all subsequent experiments (Figure 4).

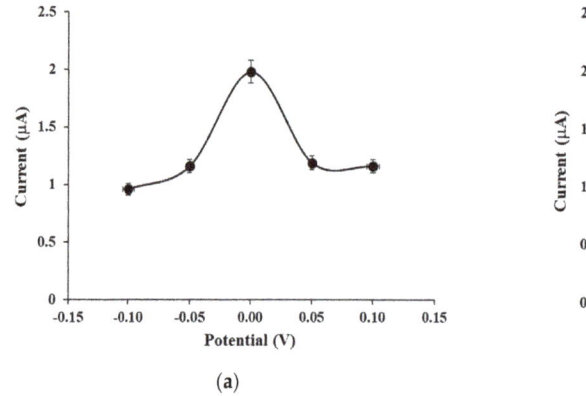
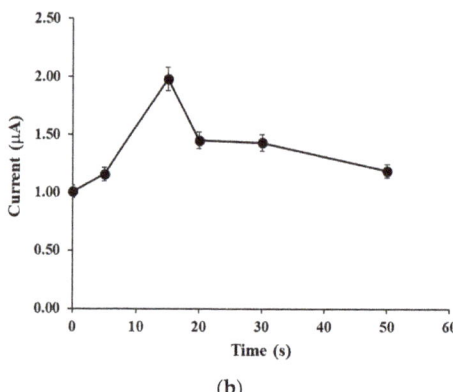

(a)  (b)

**Figure 4.** Effects of deposition potential (**a**) and (**b**) time.

*4.5. Analytical Characterization and Validation*

Under optimum deposition potential and time conditions using the AdSDPV, samples with increasing concentrations of EZE and ROS were prepared. Analytical characterization in terms of LOD and LOQ based on 3 s/m and 10 s/m, respectively, were achieved using linear curves; where m is the slope of the related calibration curves and s is the standard deviation of the peak currents of the lowest concentration of the analyte. EZE was determined in the linear range between $1.0 \times 10^{-6}$ M to $2.5 \times 10^{-5}$ M, with a LOD of $3.0 \times 10^{-7}$ M and a LOQ of $1.0 \times 10^{-6}$ M. ROS was determined in the linear range between $5 \times 10^{-6}$ M to $1.25 \times 10^{-5}$ M, with a LOD of $2.0 \times 10^{-6}$ M and a LOQ of $6.6 \times 10^{-6}$ M. For the validation of the developed method, accuracy and precision were investigated by analyzing five replicate experiments between days and within days. Relative standard deviations (RSD%) were determined to control the precision of the technique. As summarized in Table 1, the results after statistical evaluation indicate that the technique is analytically acceptable (Figure 5 and Table 1).

**Table 1.** Statistical assessment of the calibration data for determination of ROS and EZE by the AdS-DPV method in 0.1 M $H_2SO_4$ (stripping conditions: accumulation potential of 0.0 V and accumulation time of 15 s).

| | Buffer | | Serum | | Urine | |
|---|---|---|---|---|---|---|
| Compounds | EZE | ROS | EZE | ROS | EZE | ROS |
| Linearity range (M) | $1 \times 10^{-6}$–$2.5 \times 10^{-5}$ | $5 \times 10^{-6}$–$1.25 \times 10^{-5}$ | $3.0 \times 10^{-6}$–$1.0 \times 10^{-5}$ | $2.0 \times 10^{-5}$–$6.0 \times 10^{-5}$ | $3.0 \times 10^{-6}$–$1.0 \times 10^{-5}$ | $2.0 \times 10^{-5}$–$6.0 \times 10^{-5}$ |
| Slope (µA/mM) | 76.93 | 24.20 | 30.54 | 11.31 | 49.05 | 14.19 |
| Intercept (mM) | −0.03 | −0.08 | 0.03 | −0.12 | 0.09 | −0.05 |
| Determination coefficient | 0.999 | 0.999 | 0.999 | 0.999 | 0.999 | 0.999 |
| LOD (M) | $3.0 \times 10^{-7}$ | $2.0 \times 10^{-6}$ | $1.0 \times 10^{-6}$ | $1.0 \times 10^{-6}$ | $3.0 \times 10^{-7}$ | $1.0 \times 10^{-6}$ |
| LOQ (M) | $1.0 \times 10^{-6}$ | $6.6 \times 10^{-6}$ | $2.0 \times 10^{-6}$ | $4.0 \times 10^{-6}$ | $1.0 \times 10^{-6}$ | $4.0 \times 10^{-6}$ |
| Within-day Repeatability (RSD %) * | 1.20 | 1.74 | 1.45 | 1.42 | 1.68 | 1.67 |
| Between-day Repeatability (RSD %) * | 1.48 | 1.72 | 1.83 | 1.51 | 1.87 | 1.98 |

* Average of the five values.

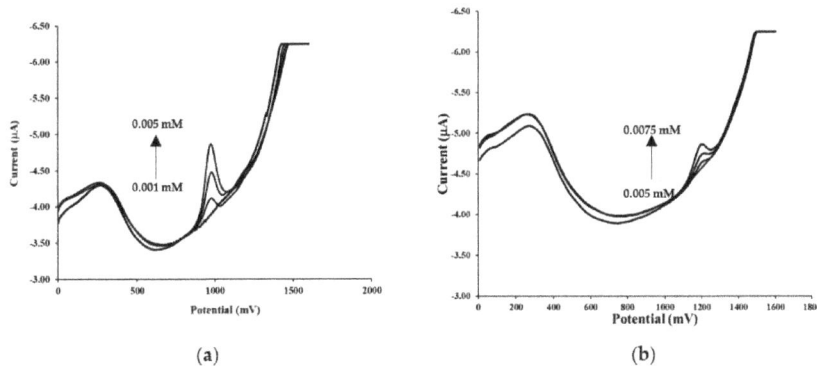

**Figure 5.** AdSDPV of sensor in various concentrations of (a) EZE and (b) ROS using the AdSDPV technique in 0.1 M $H_2SO_4$ (stripping conditions: accumulation potential of 0.0 V and accumulation time of 15 s).

*4.6. Determination of Ezetimibe and Rosuvastatin in Biological Samples*

In optimized conditions, the electrochemical method was also applied for the detection of EZE and ROS in buffer, spiked human serum, and urine samples, and reported in terms of recovery. Using the suggested method, the purified samples were used for the simultaneous determination of EZE and ROS. Recovery studies were performed by adding ROS and EZE in certain amounts to the human urine samples and serum samples by the proposed technique. The recovery studies of ROS and EZE were assessed based on the data given in Table 1. The proposed technique of RSD% and the average recovery results confirmed suitable accuracy and precision. The applicability of the developed method was indicated by constituting calibration graphs for ROS and EZE in the presence of spiked urine and serum samples. The developed technique was used for the accurate determination of ROS and EZE in biological samples without any pretreatment procedure. The outcomes of the calibration calculations and related parameters obtained in human urine and serum samples are given in Table 1. Recovery results of ROS and EZE were controlled with the corresponding calibration equations, obtained in human urine and serum samples, and found acceptable (Table 2). All results indicated the potential applicability of the developed method for evaluating human urine and serum samples.

**Table 2.** Recovery results obtained from the analysis of ROS and EZE in human urine and serum samples by AdSDPV method in 0.1 M $H_2SO_4$ (stripping conditions: accumulation potential of 0.0 V and accumulation time of 15 s).

|  | ROS | | EZE | |
|---|---|---|---|---|
|  | Human Urine | Human Serum | Human Urine | Human Serum |
| Added (mg) | 5.00 | 5.00 | 5.00 | 5.00 |
| Found (mg) * | 4.87 | 4.92 | 4.82 | 4.89 |
| Recovery (%) * | 97.4 | 98.4 | 96.4 | 97.8 |
| RSD (%) | 1.10 | 0.80 | 1.30 | 0.90 |
| Bias (%) | 2.60 | 1.60 | 3.60 | 2.20 |

* Average of the five values.

## 5. Conclusions

In this study, the electrochemical behavior of ROS and EZE was studied simultaneously for the first time. AdSDPV was used for the reliable detection of ROS and EZE in a 0.1 M $H_2SO_4$ solution with commercial deproteinated human serum samples and human urine samples using a GCE, and results were reported in terms of recovery. The developed simple and low-cost method showed high sensitivity, a low limit of detection, good repeatability, and good linearity. In the proposed technique, we monitored linear relationships varying from $1.0 \times 10^{-6}$ M to $2.5 \times 10^{-5}$ M for EZE concentrations and from $5.0 \times 10^{-6}$ M to $1.25 \times 10^{-5}$ M for ROS concentrations. LOD values were found for ROS and EZE as $3.0 \times 10^{-7}$ M and $2.0 \times 10^{-6}$ M, respectively. As is stated in the literature, for patients taking 40 mg ROS daily, the average plasma concentration of ROS ($C_{max}$) was 9.8 ng/mL (0.0098 µg/mL) [9]. Furthermore, after one dose of EZE, average EZE peak plasma concentrations ($C_{max}$) of 3.4 to 5.5 ng/mL (0.0055 µg/mL) were obtained within 4 to 12 h ($T_{max}$) [10]. These values are higher than our limit of detection value, indicating that the proposed method can be used to detect ROS and EZE in real samples.

**Author Contributions:** Conceptualization, S.K., B.U. and S.A.O.; methodology, L.K. and S.K.; software, L.K.; validation, L.K. and S.K.; formal analysis, L.K.; investigation L.K. and S.K.; resources, L.K. and S.K.; data curation, L.K.; writing—original draft preparation, L.K. and S.K.; writing—review and editing, S.K., B.U. and S.A.O.; visualization, S.A.O.; supervision, S.A.O.; project administration, S.A.O.; funding acquisition, L.K., S.K., B.U. and S.A.O. All authors have read and agreed to the published version of the manuscript.

**Funding:** This research was funded by Ankara University BAP 17B0237002.

**Institutional Review Board Statement:** Not applicable.

**Informed Consent Statement:** Not applicable.

**Data Availability Statement:** Not applicable.

**Conflicts of Interest:** The authors declare no conflict of interest.

# References

1. Chapman, M.J.; McTaggart, F. Optimizing the pharmacology of statins: Characteristics of rosuvastatin. *Atheroscler. Suppl.* **2002**, *2*, 33–37. [CrossRef]
2. Scott, L.J.; Curran, M.P.; Figgitt, D.P. Rosuvastatin: A review of its use in the management of dyslipidemia. *Am. J. Cardiovasc. Drugs* **2004**, *4*, 117–138. [CrossRef] [PubMed]
3. Zhang, R.; Li, Y.; Jiang, X.; Wang, L. Pharmacokinetics and tolerability of multiple-dose rosuvastatin: An open-label, randomized-sequence, three-way crossover trial in healthy Chinese volunteers. *Curr. Ther. Res.-Clin. Exp.* **2009**, *70*, 392–404. [PubMed]
4. Kosoglou, T.; Statkevich, P.; Johnson-Levonas, A.O.; Paolini, J.F.; Bergman, A.J.; Alton, K.B. Ezetimibe: A review of its metabolism, pharmacokinetics and drug interactions. *Clin. Pharmacokinet.* **2005**, *44*, 467–494. [CrossRef] [PubMed]
5. Phan, B.A.P.; Dayspring, T.D.; Toth, P.P. Ezetimibe therapy: Mechanism of action and clinical update. *Vasc. Health Risk Manag.* **2012**, *8*, 415–427.
6. Nakajima, N.; Miyauchi, K.; Yokoyama, T.; Ogita, M.; Miyazaki, T.; Tamura, H.; Nishino, A.; Yokoyama, K.; Okazaki, S.; Kurata, T.; et al. Effect of combination of ezetimibe and a statin on coronary plaque regression in patients with acute coronary syndrome: ZEUS trial (eZEtimibe Ultrasound Study). *IJC Metab. Endocr.* **2014**, *3*, 8–13. [CrossRef]
7. Bays, H.E.; Davidson, M.; Massaad, R.; Flaim, D.; Lowe, R.; Tershakovec, A.; Jones-Burton, C. Efficacy and Safety of Ezetimibe Plus Rosuvastatin Versus Rosuvastatin Up-Titration in Hypercholesterolemic Patients at Risk for Atherosclerotic Coronary Heart Disease. *J. Clin. Lipidol.* **2011**, *5*, 217–218. [CrossRef]
8. Nissen, S.E.; Nicholls, S.J.; Sipahi, I.; Libby, P.; Raichlen, J.S.; Ballantyne, C.M.; Davignon, J.; Erbel, R.; Fruchart, J.C.; Tardif, J.-C.; et al. Effect of Very High-Intensity Statin Therapy on Regression of Coronary Atherosclerosis. *JAMA* **2006**, *295*, 1556. [CrossRef]
9. DeGorter, M.K.; Tirona, R.G.; Schwarz, U.I.; Choi, Y.H.; Dresser, G.K.; Suskin, N.; Myers, K.; Zou, G.Y.; Iwuchukwu, O.; Wei, W.Q.; et al. Clinical and pharmacogenetic predictors of circulating atorvastatin and rosuvastatin concentrations in routine clinical care. *Circ. Cardiovasc. Genet.* **2013**, *6*, 400–408. [CrossRef]
10. Patel, J.; Sheehan, V.; Gurk-Turner, C. Ezetimibe (Zetia): A New Type of Lipid-Lowering Agent. *Baylor Univ. Med. Cent. Proc.* **2003**, *16*, 354–358. [CrossRef]
11. Ozkan, S.A.; Uslu, B. From mercury to nanosensors: Past, present and the future perspective of electrochemistry in pharmaceutical and biomedical analysis. *J. Pharm. Biomed. Anal.* **2016**, *130*, 126–140. [CrossRef] [PubMed]
12. Ozkan, S.A. *Electroanalytical Methods in Pharmaceutical Analysis and Their Validation*, 1st ed.; HNB Pub.: Palenville, NY, USA, 2012.
13. Bard, A.J.; Faulkner, L.R. *Electrochemical Methods: Fundamentals and Applications*, 2nd ed.; John Wiley & Sons: Hoboken, NJ, USA, 2001; Volume 677, ISBN 0471043729.
14. Compton, R.G.; Banks, C.E. *Understanding Voltammetry*; World Scientific Publishing Europe Ltd.: London, UK, 2010; ISBN 978-1-84816-585-4.
15. Jain, R.; Yadav, R.K.; Dwivedi, A. Square-wave adsorptive stripping voltammetric behaviour of entacapone at HMDE and its determination in the presence of surfactants. *Colloids Surf. A Physicochem. Eng. Asp.* **2010**, *359*, 25–30. [CrossRef]
16. Beludari, M.I.; Prakash, K.V.; Mohan, G.K. RP-HPLC method for simultaneous estimation of Rosuvastatin and Ezetimibe from their combination tablet dosage form. *Int. J. Chem. Anal. Sci.* **2013**, *4*, 205–209. [CrossRef]
17. Kurbanoglu, S.; Esim, O.; Ozkan, C.K.; Savaser, A.; Ozkan, Y.; Uslu, B.; Ozkan, S.A. Stability-indicating liquid chromatographic method for the simultaneous determination of rosuvastatin and ezetimibe from pharmaceuticals and biological samples. *J. Turk. Chem. Soc. Sect. A Chem.* **2020**, *7*, 865–874. [CrossRef]
18. Sharma, S.; Sharma, M.C.; Kohli, D.V.; Chaturvedi, S.C. Micellar liquid chromatographic method development for determination of rosuvastatin calcium and ezetimibe in pharmaceutical combination dosage form. *Der Pharma Chem.* **2010**, *2*, 371–377.
19. Varghese, S.J.; Ravi, T.K. Determination of rosuvastatin and ezetimibe in a combined tablet dosage form using high-performance column liquid chromatography and high-performance thin-layer chromatography. *J. AOAC Int.* **2010**, *93*, 1222–1227. [CrossRef]
20. Ashfaq, M.; Ahmad, H.; Khan, I.U.; Mustafa, G. Lc determination of rosuvastatin and ezetimibe in human Plasma. *J. Chil. Chem. Soc.* **2013**, *58*, 2177–2181. [CrossRef]
21. Pandya, C.B.; Channabasavaraj, K.P.; Shridhara, H.S. Simultaneous estimation of Rosuvastatin calcium and ezetimibe in bulk and tablet dosage form by simultaneous equation method. *Int. J. ChemTech Res.* **2010**, 2140–2144.
22. Varghese, S.J.; Ravi, T.K. Development and validation of a liquid chromatography/ mass spectrometry method for the simultaneous quantitation of rosuvastatin and ezetimibe in human plasma. *J. AOAC Int.* **2013**, *96*, 307–312. [CrossRef]
23. Bhadoriya, A.; Sanyal, M.; Shah, P.A.; Shrivastav, P.S. Simultaneous quantitation of rosuvastatin and ezetimibe in human plasma by LC–MS/MS: Pharmacokinetic study of fixed-dose formulation and separate tablets. *Biomed. Chromatogr.* **2018**, *32*, e4291. [CrossRef]

24. Laviron, E. Surface linear potential sweep voltammetry: Equation of the peaks for a reversible reaction when interactions between the adsorbed molecules are taken into account. *J. Electroanal. Chem. Interfacial Electrochem.* **1974**, *52*, 395–402. [CrossRef]
25. Karadurmus, L.; Kurbanoglu, S.; Uslu, B.; Ozkan, S.A. Differential Pulse Voltammetric Determination of Rosuvastatin Via Glassy Carbon Electrode. *Rev. Roum. Chim.* **2017**, *62*, 581–588.
26. Karadas-Bakirhan, N.; Gumustas, M.; Uslu, B.; Ozkan, S.A. Simultaneous determination of amlodipine besylate and rosuvastatin calcium in binary mixtures by voltammetric and chromatographic techniques. *Ionics (Kiel).* **2016**, *22*, 277–288. [CrossRef]
27. Silva, T.A.; Zanin, H.; Vicentini, F.C.; Corat, E.J.; Fatibello-Filho, O. Electrochemical determination of rosuvastatin calcium in pharmaceutical and human body fluid samples using a composite of vertically aligned carbon nanotubes and graphene oxide as the electrode material. *Sens. Actuators B Chem.* **2015**, *218*, 51–59. [CrossRef]
28. Kul, D.; Uslu, B.; Ozkan, S.A. Electrochemical Determination of Anti-Hyperlipidemic Drug Ezetimibe Based on its Oxidation on Solid Electrodes. *Anal. Lett.* **2011**, *44*, 1341–1357.
29. Özden, D.Ş.; Durmuş, Z.; Dinç, E. Electrochemical oxidation behavior of ezetimibe and its adsorptive stripping determination in pharmaceutical dosage forms and biological fluids. *Res. Chem. Intermed.* **2015**, *41*, 1803–1818. [CrossRef]

Article

# Determination of UV Filters in Waste Sludge Using QuEChERS Method Followed by In-Port Derivatization Coupled with GC–MS/MS

Cemile Yücel [1,2], Ilgi Karapinar [1,*], Serenay Ceren Tüzün [1], Hasan Ertaş [2] and Fatma Nil Ertaş [2]

[1] Faculty of Engineering, Environmental Engineering Department, Dokuz Eylül University, Tınaztepe Campus, Buca, İzmir 35390, Turkey
[2] Faculty of Science, Chemistry Department, Ege University, Bornova, İzmir 35040, Turkey
* Correspondence: ilgi.karapinar@deu.edu.tr

**Citation:** Yücel, C.; Karapinar, I.; Tüzün, S.C.; Ertaş, H.; Ertaş, F.N. Determination of UV Filters in Waste Sludge Using QuEChERS Method Followed by In-Port Derivatization Coupled with GC–MS/MS. *Methods Protoc.* **2022**, *5*, 92. https://doi.org/10.3390/mps5060092

Academic Editors: Victoria Samanidou, Verónica Pino and Natasa Kalogiouri

Received: 11 October 2022
Accepted: 20 November 2022
Published: 23 November 2022

**Publisher's Note:** MDPI stays neutral with regard to jurisdictional claims in published maps and institutional affiliations.

**Copyright:** © 2022 by the authors. Licensee MDPI, Basel, Switzerland. This article is an open access article distributed under the terms and conditions of the Creative Commons Attribution (CC BY) license (https://creativecommons.org/licenses/by/4.0/).

**Abstract:** UV filters (UVFs) are widely used in personal care and in industrial products for protection against photodegradation. In recent years, their potential toxicological and environmental effects have received growing attention. Due to their excessive use, their residue levels in the environment are gradually increasing and they tend to accumulate on biological wastewater treatment sludge. The utilization of sludge as fertilizer could be one of the main routes of UVF contamination in the environment. Therefore, the development of a reliable and sensitive method of analyzing their trace level residues in waste sludge samples is of great importance. The success of the method largely depends on the sample preparation technique in such complex matrices. This study presents a rapid, sensitive and green analysis method for eight UVFs in sludge samples, selected for their rather low no-observed-effect concentrations (NOEC). For this purpose, the QuEChERS methodology was coupled with in-port derivatization for subsequent detection of the targeted UVFs via GC–MS/MS. The analysis time was substantially shortened using this method, and reagent utilization was also reduced. The method was validated in the sludge samples, and high recovery (66–123%) and low RSD values (<25.6%) were obtained. In addition, major contributing uncertainty sources and expanded uncertainties were determined.

**Keywords:** UVFs; QuEChERS; in-port derivatization; waste sludge

## 1. Introduction

UVFs are the general name for the chemical group that absorbs ultraviolet light, through which the adverse effects of UV light are eliminated. UVFs can be grouped as inorganic, such as $TiO_2$ or $ZnO$ [1], and organic, mostly used in personal care products, as well as in plastics, automobile paints and rubber industries to increase resistance towards UV light degradation [2]. These compounds can reach to water bodies through industrial and domestic effluents. Since their utilization in personal care products and industrial applications are widespread, the UVF residue levels in the environment are increasing extensively [3].

Organic UVFs are a wide range of compounds that differ in their structures and properties. These highly persistent compounds in the environment tend to accumulate in water [3–5], suspended particles, soil [6], sediment [7] and sludges [8–10]. Organic UVFs can also accumulate in biota and disrupt the endocrine systems of aquatic organisms by increasing their estrogen-induced cell proliferation [11]. It has been clearly stated that UVFs affect different hormonal targets as well as estrogenic activity in mammals and fish, and are therefore known as endocrine-disrupting chemicals (EDCs) [12,13]. Consequently, one of these compounds has recently been included in the Watch List by The European Water Framework Directive as a future priority pollutant [14]. Although the use of these compounds in cosmetics is legally restricted [15], there is no limit to their concentration in water and sewage sludge matrices. However, studies on the presence of UVFs in waste sludge have revealed that their concentration range from a few to thousands of µg/g in dry mass [16]. The use of this sludge as fertilizer is another concern due to the widespread distribution of UVF residues in agricultural soils and the possible contamination of crops.

Since their concentrations are very low, an appropriate preparation technique must be applied to the samples to isolate and preconcentrate these filters. Current trends in the determination of organic UVFs in environmental water samples based on microextraction techniques have been reviewed [3,17,18]. However, their determination in sludge samples faces difficulties due to the complexity of the matrices and their impact on the environment. To date, one of the techniques used for UVF extraction from sewage sludge comprises liquid–liquid extraction (LLE) followed by solid-phase extraction (SPE) [19–22] where the excessive use of solvent is the main drawback of the method. The other technique is pressurized liquid extraction (PLE), applied alone or coupled with SPE [23–29] or gel permeation chromatography (GPC) [30]. Despite the advantage of using a smaller solvent volume in the PLE technique, special equipment is required to conduct this method.

In the last few decades, method development studies have been devoted to modern sample preparation techniques based on shorter analytical periods and the minimization of organic solvent utilization for a wide range of pollutants. One of these modern techniques, called QuEChERS (which is an acronymic for quick, easy, cheap, effective, rugged and safe) has been well-established to improve laboratory efficiency and throughput. The main advantage of this method is its small quantities of solvent utilization. The method has already been tested for UVFs in human milk [31] and seafood samples [32], followed by liquid chromatography systems coupled with mass detectors (LC-MS) and sludge samples, and then, by detection using gas chromatography (GC–MS) [16,21–23]. LC-MS systems are preferred for the analysis of polar UVFs, and no derivatization step is required in accordance with the physicochemical properties of UVFs [33–36]. GC–MS has been used for rather non-polar and volatile UV filters. Although the GC method provides high separation efficiency, high selectivity and good sensitivity, it displays some disadvantages such as the derivatization step, in which more reagents, greater reaction time and more labor are required. Fortunately, the in-port derivatization technique is a practical and environmentally friendly solution to these issues wherein the reaction takes place in the injection block rather than in an off-line interaction. The analysis time can be substantially reduced using this method and the selectivity can be improved since the technique enables extra purification for the matrix effect.

In the present study, we aimed to detect eight UVFs in sludge samples. Benzophenone-3 (BP-3), 3-benzylidene camphor (3BC), 2-ethyl hexyl-4-(dimethyl amino) benzoate (EDP), 2-ethyl-hexyl-4-trimethoxy cinnamate (EHMC), ethylhexyl salicylate (EHS), homosalate (HMS), isoamyl p-methoxycinnamate (IAMC) and 4-methylbenzylidene camphor (4-MBC) were selected according to their no-observed-effect concentration (NOEC) values. Their chemical structures and NOEC limits are presented in Table 1. Considering the hydrophobicity (log$K_{ow}$: 3.79–6.16) of the analytes, the QuEChERS method was adopted; however, due to the low volatility of these analytes along with their weak acidity (p$K_a$: 7.56–8.13), the analytes were derivatized using an in-port silylation technique prior to subsequent quantification using the GC–MS/MS system. To the best of our knowledge, this is the first report in the literature about the combination of QuEChERS with in-port derivatization followed by GC–MS/MS for a wide variety of UVF analyses in waste sludge. This method extensively decreases solvent or chemical consumption and shortens the analysis time compared to off-line derivatization. The method validation parameters, major contributing uncertainty sources and expanded uncertainties were determined.

Table 1. Chemical Structures, IUPAC names and NOEC concentration of selected UVFs.

| | |
|---|---|
| 3-BC: 3-benzyline camphor (3-benzylidene-1,7,7-trimethylbicyclo [2.2.1] heptan-2-one) NOEC: 0.022 mg L$^{-1}$ | BP-3: benzophenone-3 (2-hydroxy-4-methoxyphenyl)-phenyl methanone NOEC > 0.01 mg L$^{-1}$ |
| EDP: 2-ethyl hexyl-4-(dimethylamino) benzoate NOEC: 0.012 mg L$^{-1}$ | EHMC: 2-ethyl-hexyl-4-trimethoxy cinnamate NOEC: 0.003 mg L$^{-1}$ |
| EHS: 2-ethylhexyl 2-hydroxybenzoate (ethylhexylsalicylate) NOEC: 0.008 mg L$^{-1}$ | HMS: 3,3,5-trimethylcyclohexyl 2-hydroxybenzoate (Homosalate) NOEC: 0.005 mg L$^{-1}$ |
| IAMC: isoamyl p-methoxy cinnamate (3-methylbutyl (2E)-3-(4-methoxyphenyl) acrylate) NOEC: 0.013 mg L$^{-1}$ | 4-MBC: 4-methylbenzylidene camphor NOEC: 0.008 mg L$^{-1}$ |

## 2. Materials and Method

### 2.1. Reagents and Standard Solutions

The chemical standards of 2-ethylhexyl salicylate (EHS), 2-hydroxy-4-methoxybenzophenone (benzophenone-3, BP-3) >98%, 3,3,5-trimethylciclohexylsalicylate (homosalate, HMS) >98%, 4-methylbenzylidene camphor (4-MBC) 99%, 2-ethyl-hexyl-4-trimethoxy cinnamate (EHMC) 99%, and 2-ethyl hexyl-4-(dimethylamino) benzoate (EDP) 99% were obtained from Dr. Ehrenstorfer (Augsburg, Germany). 3-benzylidene camphor (3BC) 99%, isoamyl p-methoxycinnamate (IAMC) and % were acquired from Sigma-Aldrich (St. Louis, MO, USA), Toronto Research Chemicals (Toronto, ON, Canada) and Chemservice (West Chester, PA, USA), respectively.

LC-MS-grade methanol and acetonitrile, GC–MS-grade ethyl acetate (EtAC) and acetone, sodium chloride, sodium sulfate and orthophosphoric acid were purchased from Merck (Darmstadt, Germany). Bis(trimethylsilyl) trifluoroacetamide with 1% trimethylchlorosilane (BSTFA + TMCS; 99:1, $v/v$) from Macherey-Nagel was used as derivatization reagent. This reagent was stable for only 3 h during analysis. 2-dodecanol, 2-dodecanone, and 1-undecanol from Sigma were used as extraction solvents. Anhydrous magnesium sulphate ($MgSO_4$), primary and secondary amine exchange bonded silica sorbent (PSA) and octadecylsilan (C18) were obtained from Supelco (Bellefonte, PA, USA) for the extraction step.

The stock solutions of individual UVFs (1000 µg mL$^{-1}$) and mixtures of UVFs (50 µg mL$^{-1}$) were prepared in ethyl acetate to optimize the injection port derivatization conditions in methanol to validate the QuEChERS method. These solutions of 1000 µg mL$^{-1}$ and 50 µg mL$^{-1}$ were stable for about 5 months and five days at −20 °C, respectively. The working aqueous solutions were prepared daily from standards in methanol at different concentrations using ultrapure water and environmental water. The sludge samples were collected from a domestic wastewater treatment plant in Izmir, Turkey. Sludges were air-dried and stored in the dark at −20 °C until analysis.

## 2.2. Sample Extraction

The procedure for extracting UVFs from sludge is a modified version of the method reported previously [16]. As shown in Figure 1, a 0.5 g sewage sludge sample was dried at room temperature and a spiked sample was transferred into a conical-bottom 15 mL polypropylene tube containing 10 mL of ACN, vortexed for 2.5 min, and then, left in an ultrasonic bath (J.P. Selecta, Barcelona, Spain) for 15 min. Then, the organic phase was separated via centrifugation at 3500 rpm for 15 min and transferred to a conical-bottom 15 mL polypropylene tube containing 500 mg $MgSO_4$, 410 mg C18 and 315 mg PSA. The extract was then vortexed for 2.5 min and centrifuged for 15 min. After the supernatant was transferred to a 10 mL tube, it was evaporated to dryness under $N_2$ gas. Then, the extract was dissolved in 1000 µL EtAC, instead of hexane as performed by Ramos et al. [16], for the in-port derivatization step as previously applied to a surface water sample by our research group [37]. The extract dissolved in EtAC was filtered through a 13 mm, 0.22 µm PTFE filter, and then, it was transferred to a 1.5 mL amber vial. Finally, 2 µL of BSTFA and 2 µL of extract were derivatized in the injection port using the sandwich technique. In this technique, two aliquots of 2 µL BSTFA and 2 µL extract in EtAc, separated with an air gap, are drawn into the microsyringe of a PAL autosampler, and then, injected to the GC–MS/MS system.

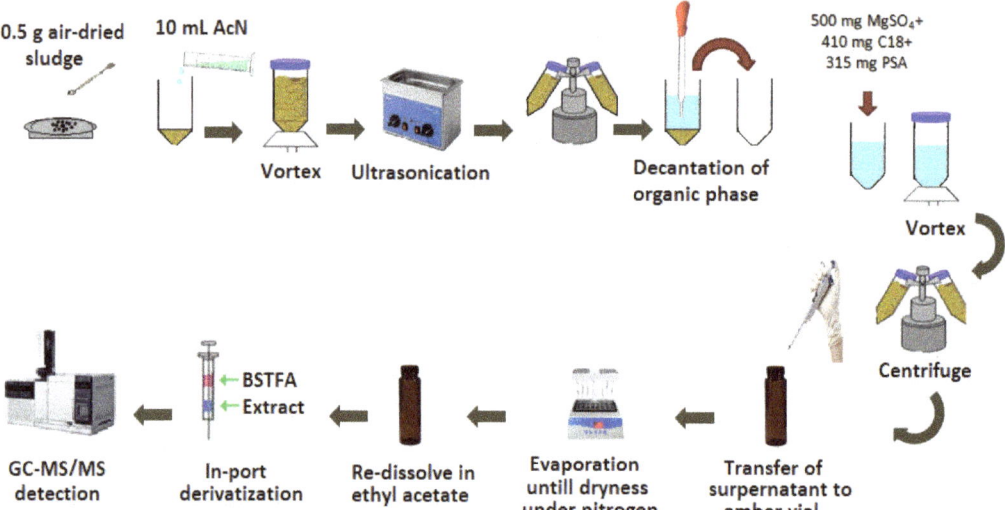

**Figure 1.** Schematic illustration of the whole procedure.

## 2.3. GC–MS/MS Analysis

GC–MS/MS analysis was performed using an Agilent 7890 B gas chromatograph coupled with a triple quadrupole mass spectrometer (MS 7000C) and a PAL autosampler (GC Sampler 80). For derivatization, the injection port temperature was held at 70 °C for 3 min, then, increased to 300 °C at a rate of 400 °C min$^{-1}$ in splitless mode with a purge-off time of 4.5 min. The oven temperature started at 70 °C for 4 min, increased to 180 °C at 25 °C min$^{-1}$, then, increased to 230 °C at a rate of 5 °C min$^{-1}$ and to 300 °C at a rate of 25 °C min$^{-1}$; it was held for 10 min at this temperature. The injector was operated using programmed temperature evaporation (PTV).

Separation was performed on 5% phenyl-arylene/95%-dimethylpolysiloxane HP-5MS (30 m × 0.25 mm i.d. 0.25 μm film thickness) supplied by Phenomenex (Torrance, CA, USA). Helium (99.999% purity) was used as the carrier gas at a constant flow of 1 mL min$^{-1}$. In MS/MS analysis, the temperatures of the ion source and the transfer line were 280 and 300 °C, respectively. The multiple reaction monitoring (MRM) technique was applied, and electron ionization (EI) mode was used. The retention times ($R_t$) obtained, the optimized MRM transitions and the collision energies (CE) for each UVF are given in Table 2. The bold parent and product ions in the table show the quantification transitions.

**Table 2.** Experimental GC–MS/MS parameters of UVFs.

| UVFs | $R_t$ (min) | MW (g mol$^{-1}$) | Parent Ions (*m/z*) | Product Ions (*m/z*) | CE (eV) |
|---|---|---|---|---|---|
| 3-BC | 14.443 | 240.35 | **240.0** | 149.2 | 5 |
|  |  |  |  | 225.1 | 9 |
|  |  |  |  | 92.10 | 12 |
| EHS | 14.452 | 250.34 | **195.0** | 177.0 | 15 |
|  |  |  |  | 159.0 | 25 |
|  |  |  |  | 75.00 | 30 |
| HMS | 15.610 | 262.36 | **195.0** | 177.0 | 15 |
|  |  |  |  | 159.0 | 25 |
|  |  |  |  | 75.00 | 27 |
| IAMC | 15.737 | 248.32 | **178.1** | **161.1** | 15 |
|  |  |  | 161.0 | 133.0 | 10 |
| 4-MBC | 16.120 | 254.37 | **254.0** | 239.0 | 15 |
|  |  |  |  | 105.0 | 25 |
| BP-3 | 16.527 | 228.25 | **285.0** | 242.0 | 25 |
| EDP | 19.014 | 277.41 | **277.0** | **164.9** | 10 |
|  |  |  | 148.0 | 104.2 | 30 |
|  |  |  | 165.0 | 148.6 | 32 |
| EHMC | 19.501 | 290.41 | **161.0** | **133.1** | 8 |
|  |  |  | 178.0 | 133.1 | 22 |
|  |  |  | 290.0 | 178.1 | 6 |

## 2.4. Validation Studies

The matrix match method was used for the calibration curves. In this method, the curves were constructed by subtracting the peak area values of the real sludge sample from the spiked extract and plotting against the concentration of the UV filter added into the real sludge sample.

The linear range, intra-day and inter-day repeatability, Limit of Detection (LOD), Limit of Quantitation (LOQ) and recovery parameters were determined, and measurement uncertainties of the method applied for UVFs were calculated. For the linearity of UVFs, sludge samples were spiked with 40, 80, 200, 600 and 1200 ng g$^{-1}$ UVF standards. The extracts were evaporated to dryness under nitrogen gas, and they were diluted to 1000 μL using ethyl acetate. The LOQ and LOD were calculated according to S/N = 10 and

S/N = 3, respectively. Intra-day and inter-day repeatability studies were performed at low (80 ng g$^{-1}$), medium (300 ng g$^{-1}$), and high (600 ng g$^{-1}$) concentrations, with 3 replicates.

The selected test material was analyzed repeatedly under different conditions such as on different days, using different analysts and different equipment, etc. The total variation in the whole cluster can be represented as the combination of variances (s$^2$) between (S$_{between}$) and within groups (S$_r$). The repeatability (intra-day) and intermediate precision (inter-day) values were calculated via ANOVA [38]. The standard deviation of S$_r$ was calculated by taking the square root of the within-group mean square term, as shown in Equation (1), and the contribution of the grouping factor to the total variation was obtained from Equation (2). In Equations (1) and (2), MS$_w$ is the within-group mean square term and $MS_b$ is the between-group mean square term.

Then, intermediate precision (S$_I$) was calculated by combining the within-group and between-group variance components, as shown in Equation (3).

$$S_r = \sqrt{MS_w} \quad (1)$$

$$S_{between} = \sqrt{\frac{MS_b - MS_w}{n}} \quad (2)$$

$$S_I = \sqrt{S_r^2 + S_{between}^2} \quad (3)$$

The intra-day (n = 3) and inter-day (n = 2) relative standard deviations (RSD%) of the QuEChERS followed by GC–MS/MS were obtained using spiked solutions of the analytes at different concentration levels. According to the 2015/1787 directive, if the RSD% value of the applied method is less than 25%, the precision of this method is acceptable for organic compounds.

Recovery studies to determine the accuracy of the method were carried out by adding the UVF standards to the sewage sludge samples at concentrations of 80, 300 and 600 ng g$^{-1}$. The recovery percentages were calculated from Equation (4) where, $Cpre - Ext$ and $Cpost - Ext$ are the concentrations of sludges in which analytes were added before and after extraction, respectively. $Csample$ is the concentration of UVFs in sludges without the addition of the analyte.

$$\text{Recovery \%} = \frac{Cpre - Ext - Csample}{Cpost - Ext - Csample} \times 100 \quad (4)$$

The measurement uncertainty is a parameter that was included with the measured result and characterizes the distribution of values that can correspond to the measurand. Knowing the uncertainty means increased confidence in the accuracy of the measurement result. This value is very important in comparing the measurement results of two different methods and deciding whether the results are within the defined limits. Measurement uncertainty consists of many components. Some of these components are derived from the statistical distribution of the results of repeated measurement series to obtain the standard deviations. Combined standard uncertainty ($u(c)$) is the standard uncertainty that considers contributions from all important uncertainty sources by combining the relevant uncertainty components, and it was calculated as shown in Equation (5). The expanded uncertainty provides the range of an analyte concentration believed to be spread at a higher confidence level. The expanded uncertainty ($U$) is calculated by multiplying the combined standard uncertainty by "$k$", which is equal to 2 at 95% confidence level [39]. In this study, the uncertainty sources were defined first; then, the uncertainty of each parameter was determined, and finally, the combined $Uc(y)$ and expanded uncertainties ($U$) were calculated.

$$u(c) = \sqrt{u_{calibration}^2 + u_{SI}^2 + u_{Recovery}^2} \quad (5)$$

## 3. Results and Discussion

### 3.1. Optimization Studies

For the extraction of targeted UVFs from sludge samples, the QuEChERS methodology was adopted for the in-port derivatization and the extract obtained was dissolved in 1000 µL ethyl acetate (EtAc) instead of hexane. In a previous study carried out in this lab, in-port derivatization conditions were optimized for a wide range of UVFs extracted from surface water via vortex-assisted dispersive liquid–liquid microextraction based on the solidified floating organic droplet (VA-DLLME-SFOD) technique. It was determined that the injection temperature was a statistically significant factor and the optimal temperature was determined to be 260 °C [37]. The effect of injection temperature for the studied UVFs in sludge samples was further studied to see any deviation from the optimal conditions determined. The temperature varied between 260–320 °C, and the peak areas are given in Figure 2.

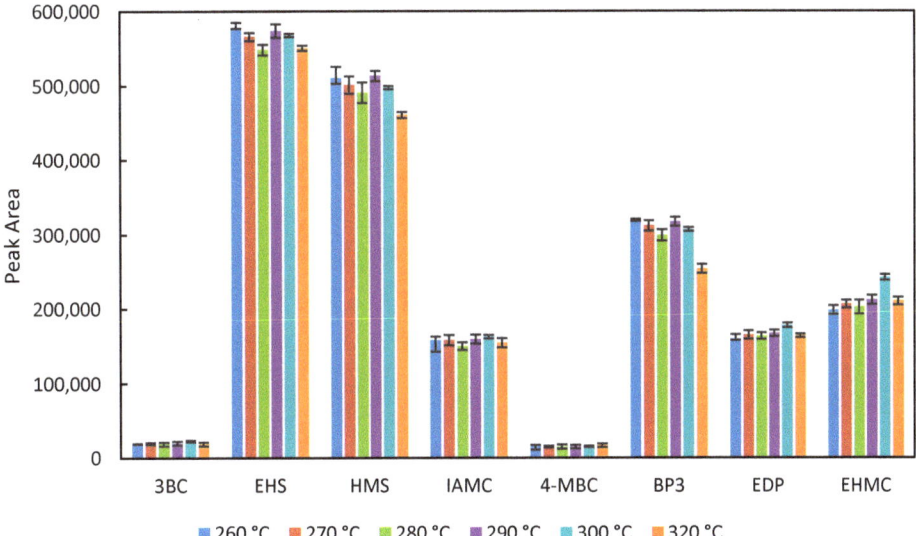

**Figure 2.** The effect of injection temperature on peak areas of UVFs.

As can be deduced from the figure, even though the mean peak areas of the UVFs are not substantially different from each other for the studied injection temperatures, a slight decrease was observed for 300 °C. However, this injection temperature was chosen for further studies since the carry-over effect occurred in samples up to 290 °C. The sharp peaks in UVFs with injection port derivatization, after QuEChERS, of the spiked sludge sample were obtained, as shown in Figure 3.

### 3.2. Validation Studies

The linearity and linear range of a method should unequivocally be determined for any analytical method. It is a fact that the lowest concentration of the calibration curve should be very close to the LOQ value for the accurate analysis of analytes with known precision at an LOQ-level concentration. Table 3 depicts the linear equations for the working range of 40–1200 ng g$^{-1}$ with correlation coefficients close to unity ($R^2 > 0.9970$).

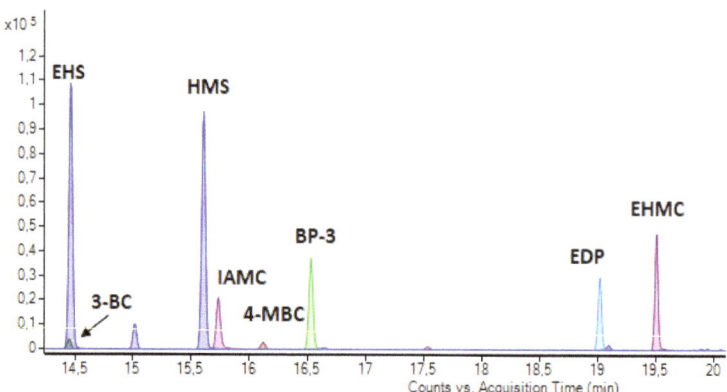

**Figure 3.** Chromatograms of the UV filters in sewage sludge spiked with UVFs at 300 ng g$^{-1}$.

**Table 3.** Analytical merits of the GC–MS/MS method coupled with QuEChERS for UVF determination in sludge samples.

| Analyte | a * | b * | R$^2$ | LOD (ng g$^{-1}$) | LOQ (ng g$^{-1}$) |
|---|---|---|---|---|---|
| 3-BC | 30.154 | −591.4 | 0.9991 | 10.1 | 39.6 |
| EHS | 888.02 | −23,900 | 0.9987 | 10.0 | 33.3 |
| HMS | 515.64 | −11,424 | 0.9984 | 9.90 | 33.0 |
| IAMC | 170.2 | −285.3 | 0.9996 | 9.50 | 31.8 |
| 4-MBC | 24.748 | −159.2 | 0.9997 | 8.00 | 26.7 |
| BP−3 | 276.81 | −10,099 | 0.9980 | 4.30 | 14.1 |
| EDP | 214.00 | −1290 | 0.9996 | 12.1 | 40.2 |
| EHMC | 204.46 | −3367 | 0.9970 | 8.00 | 26.8 |

* Y = ax + b.

LOQ is defined as the lowest concentration that can be measured with acceptable precision (20% RSD) and accuracy [40–45]. Three different methods are used to determine the LOQ value. The most common approach for LOQ calculation in chromatographic analysis is the signal-to-noise ratio (S/N) [45]. This ratio can be defined as the difference between the height of the analyte peak (signal) and the highest and lowest points of the baseline (noise) in each area around the signal. For LOQ, S/N usually needs to be at least equal to 10. In the second method, a specific calibration curve should be studied using samples containing an analyte in the range of the LOQ. The residual standard deviation of a regression line or the standard deviation of the y-intercepts of regression lines may be used as the standard deviation [39]. The third approach is the concentration corresponding to a response 10 times greater than the SD of the analysis at the minimum concentration [37,41,45].

The LOD and LOQ values were calculated based on the S/N = 3 approach and S/N = 10, respectively. The LOQ values calculated using this method were found to be in agreement with previous studies on the analysis of UVFs in sludge [17,24,26]. In-port derivatization provides lower detection limits by converting polar analytes to more volatile compounds, as reported earlier [46–48]. For most of the compounds, the LOQ is close to or less than the lowest concentration level of the calibration curve. The RSD% values obtained in this study are below 25% as seen in Table 4.

**Table 4.** Intra-day (n = 3) and inter-day (n = 2) relative standard deviations (RSD%) calculated for the UVFs.

| UVFs | Intra-Day RSD% (n = 3) | | | Inter-Day RSD% (n = 2) | | |
|---|---|---|---|---|---|---|
|  | 80 ng g$^{-1}$ | 300 ng g$^{-1}$ | 600 ng g$^{-1}$ | 80 ng g$^{-1}$ | 300 ng g$^{-1}$ | 600 ng g$^{-1}$ |
| 3-BC | 4.3 | 8.7 | 3.2 | 10.8 | 9.1 | 7.1 |
| EHS | 9.9 | 9.7 | 17.2 | 10.9 | 10.9 | 19.3 |
| HMS | 13.1 | 12.1 | 21.9 | 19.8 | 13.7 | 25.6 |
| IAMC | 5.2 | 8.2 | 3.4 | 15.1 | 17.9 | 19.8 |
| 4-MBC | 4.0 | 8.1 | 3.2 | 10.1 | 10.7 | 7.2 |
| BP-3 | 16.2 | 11.6 | 13.6 | 17.7 | 13.7 | 15.5 |
| EDP | 4.9 | 11.2 | 2.3 | 15.0 | 14.4 | 15.4 |
| EHMC | 9.2 | 12.9 | 3.4 | 22.6 | 15.9 | 18.8 |

The recoveries were in the range of 66%–23% as depicted in Table 5. Although, the recoveries are almost the same as the results obtained in other studies [17,24,26,31], our method provides certain advantages such as the lack of special apparatus and comparatively shorter analysis times.

**Table 5.** Recoveries of UVFs from sludge samples spiked with different concentrations.

| UVFs | % Recovery ± RSD (n = 3) | | |
|---|---|---|---|
|  | 80 ng g$^{-1}$ | 300 ng g$^{-1}$ | 600 ng g$^{-1}$ |
| 3-BC | 115 ± 4.49 | 93 ± 10.1 | 98 ± 4.03 |
| EHS | 113 ± 6.83 | 98 ± 10.2 | 96 ± 7.34 |
| HMS | 87 ± 11.1 | 109 ± 12.5 | 104 ± 25.8 |
| IAMC | 121 ± 2.70 | 118 ± 8.80 | 113 ± 4.70 |
| 4-MBC | 97 ± 2.35 | 75 ± 9.12 | 66 ± 3.05 |
| BP-3 | 88 ± 18.1 | 87 ± 14.9 | 66 ± 14.7 |
| EDP | 106 ± 2.25 | 103 ± 12.1 | 103 ± 3.21 |
| EHMC | 123 ± 7.49 | 108 ± 14.4 | 107 ± 4.43 |

### 3.3. Measurement Uncertainty

Here, the main uncertainty components such as $u_{calibration}$, $u_{SI}$ and $u_{Recovery}$ were calculated for the analysis of two different UVF concentrations, as given in Table 6. The uncertainty of the QuEChERS-GC–MS/MS method for UVFs is between 13.2–47.4% at a concentration of 300 ng g$^{-1}$ and between 6.9–43.6% at a concentration of 600 ng g$^{-1}$, which are quite satisfactory in such a complex matrix. Uncertainties regarding the recovery and calibration curve were the largest source contributing to measurement uncertainty.

**Table 6.** Expanded uncertainties of QuEChERS followed by the in-port derivatization method for studied UVFs.

| UV Filter | Description | Value (ng g$^{-1}$)x | | Standard Uncertainty u(x) | | Relative Standard Uncertainty u(x) | |
|---|---|---|---|---|---|---|---|
|  |  | 300 | 600 | 300 | 600 | 300 | 600 |
| EHS | Repeatability | 1 | 1 | 0.0559 | 0.0860 | 0.0559 | 0.0860 |
|  | Bias (recovery) | 0.8700 | 0.9948 | 0.0336 | 0.0582 | 0.0385 | 0.0585 |
|  | Calibration | 300 | 600 | 14.855 | 15.075 | 0.0495 | 0.0251 |
|  | u(c) |  |  |  |  | 0.0841 | 0.1070 |
|  | Expanded U(x) |  |  |  |  | 0.1681 | 0.2139 |

Table 6. Cont.

| UV Filter | Description | Value (ng g$^{-1}$)x | | Standard Uncertainty $u(x)$ | | Relative Standard Uncertainty $u(x)$ | |
|---|---|---|---|---|---|---|---|
| | | 300 | 600 | 300 | 600 | 300 | 600 |
| HMS | Repeatability | 1 | 1 | 0.0696 | 0.1267 | 0.0696 | 0.12673 |
| | Bias (recovery) | 0.8367 | 0.9183 | 0.1719 | 0.0939 | 0.2054 | 0.1022 |
| | Calibration | 300 | 600 | 16.826 | 17.0757 | 0.0561 | 0.0284 |
| | $u(c)$ | | | | | 0.2241 | 0.1653 |
| | Expanded $U(x)$ | | | | | 0.4481 | 0.3306 |
| 3-BC | Repeatability | 1 | 1 | 0.0502 | 0.0182 | 0.0502 | 0.0182 |
| | Bias (recovery) | 0.9064 | 0.9294 | 0.0361 | 0.0268 | 0.0398 | 0.0288 |
| | Calibration | 300 | 600 | 12.634 | 12.822 | 0.0421 | 0.0214 |
| | $u(c)$ | | | | | 0.0767 | 0.0403 |
| | Expanded $U(x)$ | | | | | 0.1534 | 0.0805 |
| IAMC | Repeatability | 1 | 1 | 0.0473 | 0.0198 | 0.0473 | 0.0198 |
| | Bias (recovery) | 0.9653 | 0.9527 | 0.0677 | 0.0756 | 0.0702 | 0.0793 |
| | Calibration | 300 | 600 | 8.2934 | 8.2934 | 0.0276 | 0.0138 |
| | $u(c)$ | | | | | 0.0890 | 0.0829 |
| | Expanded $U(x)$ | | | | | 0.1780 | 0.1658 |
| 4-MBC | Repeatability | 1.0000 | 1 | 0.0465 | 0.0185 | 0.0465 | 0.0185 |
| | Bias (recovery) | 0.6423 | 0.6257 | 0.0262 | 0.0167 | 0.0408 | 0.0266 |
| | Calibration | 300 | 600 | 6.6834 | 6.7825 | 0.0223 | 0.0113 |
| | $u(c)$ | | | | | 0.0657 | 0.0343 |
| | Expanded $U(x)$ | | | | | 0.1315 | 0.0687 |
| BP-3 | Repeatability | 1 | 1 | 0.0670 | 0.0784 | 0.0670 | 0.0784 |
| | Bias (recovery) | 0.8426 | 0.6435 | 0.0507 | 0.0342 | 0.0602 | 0.0531 |
| | Calibration | 300 | 600 | 18.7049 | 18.9821 | 0.0623 | 0.0316 |
| | $u(c)$ | | | | | 0.1096 | 0.0998 |
| | Expanded $U(x)$ | | | | | 0.2191 | 0.1997 |
| EDP | Repeatability | 1 | 1 | 0.0645 | 0.0135 | 0.0645 | 0.0135 |
| | Bias (recovery) | 0.8690 | 0.9033 | 0.0477 | 0.0505 | 0.0548 | 0.0559 |
| | Calibration | 300 | 600 | 14.1641 | 14.3741 | 0.0472 | 0.0240 |
| | $u(c)$ | | | | | 0.0970 | 0.0623 |
| | Expanded $U(x)$ | | | | | 0.1939 | 0.1246 |
| EHMC | Repeatability | 1 | 1 | 0.0747 | 0.0193 | 0.0747 | 0.0193 |
| | Bias (recovery) | 0.7552 | 56,847 | 0.1589 | 1.2119 | 0.2104 | 0.2132 |
| | Calibration | 300 | 600 | 23.9792 | 24.3341 | 0.0799 | 0.0406 |
| | $u(c)$ | | | | | 0.2371 | 0.2179 |
| | Expanded $U(x)$ | | | | | 0.4743 | 0.4357 |

### 3.4. Comparison with Other Methods

The recovery values obtained for all analytes are similar to those of studies with laborious and expensive techniques. In some studies, recovery values at a single concentration were determined [21,22,27,36], while in others, recovery values at different concentrations were investigated [16,19,23,33]. Our results indicated that for 4-MBC, only at a high concentration (600 ng g$^{-1}$) was the recovery value lower than that of the reported ones. Fortunately, good recovery values were obtained at lower concentrations. In addition, to the best of our knowledge, method validation for the analysis of 3-BC, EHS, HMS and IAMC in sludge was attained successfully. Analytes (EHS, HMS, BP-3) were derivatized using the in-port derivatization technique, which is fast, reliable and eco-friendly without being affected by the complexity of the matrix.

The LOD values of the selected UVFs in the spiked sludge sample ranged from 8 ng g$^{-1}$ to 12.1 ng g$^{-1}$. These limits are quite successful when compared to other studies (Table 7). The LOD values obtained for UVFs were lower than previous studies for 4-MBC, EHMC [23,27], BP-3 [23,36] and EDP [16]. On the other hand, the LOD values for 4-MBC [21,22,33], EHMC [16,21,22], BP-3 [19,27,33] and EDP [23] were higher.

Table 7. Comparison of QuEChERS method with previous microextraction studies in sludge samples.

| UV Filters | Extraction Method | Instrumental Method | LOD (ng g$^{-1}$) | Recovery (%) | RSD% | References |
|---|---|---|---|---|---|---|
| 4-MBC | | | 12 | 102 | 16–10 | |
| OC | | | 18 | 70 | 4–9 | |
| EHMC | | | 19 | 90 | 5–10 | |
| ODP | PLE | UPLC-MS/MS | 0.2 | 85 | 7 | [27] |
| BP-3 | | | 1.0 | 70 | 5–9 | |
| BP-1 | | | 60 | 30 | 9–14 | |
| 4HB | | | 5.0 | 95 | 4–11 | |
| 4DHB | | | 5.0 | 96 | 3–6 | |
| EHS | | | 17 * | 95–101 | 7 | |
| HMS | | | 34 * | 78–96 | 5–6 | |
| IAMC | | | 34 * | 80–107 | 4–6 | |
| BP-3 | PLE + SPE | GC–MS | 61 * | 89–106 | 6–11 | [23] |
| 4-MBC | | | 26 * | 79–86 | 4–5 | |
| EDP | | | 22 * | 88–93 | 6–7 | |
| EHMC | | | 24 * | 73–90 | 5 | |
| OC | | | 33 * | 84–85 | 5–12 | |
| BP-1 | | | 2.5 * | 74 ± 9 | 9 | |
| BP-2 | | | 2.5 * | 99 ± 11 | 11 | |
| BP-3, | PLE | LC-MS/MS | 25 * | 104 ± 14 | 14 | [36] |
| BP-4 | | | 5 * | 114 ± 28 | 28 | |
| PBSA | | | 5 * | 118 ± 19 | 19 | |
| BP-1, | | | 0.41 | | | |
| BP-2 | | | 0.67 | | | |
| BP-3 | | | 0.67 | | | |
| BP-8 | LLE + SPE | LC-MS/MS | 0.41 | 38.3–116 | 3.14–13.8 | [19] |
| 1H-BT | | | 0.67 | | | |
| 5Me-1H-BT | | | 0.67 | | | |
| TBHPBT | | | 0.1 | | | |
| 4-OH-HB | | | 0.41 | | | |
| 4-MBC | | | 4 | 94.6 | 13.1 | |
| EHMC | LLE-SPE | GC-MS | 3 | 101.2 | 10.5 | [21] |
| OC | | | 6 | 87.5 | 7.5 | |
| 4-MBC | | | 4 | 95 | 2 | |
| EHMC | LLE-SPE | GC–MS | 3 | 101 | 13 | [22] |
| OC | | | 6 | 87 | 7 | |
| BP | | | 0.3 | 63–82 | 0.1–1.0 | |
| BP-3 | QuEChERS | UPLC-MS/MS | 0.3 | 60–86 | 0.2–0.5 | [33] |
| 4-MBC | | | 0.6 | 86–95 | 0.1–6.0 | |
| BP | | | 26 | 92–101 | 2–6 | |
| 4-MBC | | | 59 | 85–88 | 2–7 | |
| EDP | QuEChERS | GC-MS/MS | 31 | 82–86 | 1–2 | [16] |
| EHMC | | | 5 | 113–125 | 1–5 | |
| OC | | | 6 | 81–94 | 3–5 | |
| 3-BC | | | 10.1 | 93–115 | 3.2–10.8 | |
| EHS | | | 10.0 | 96–113 | 9.7–19.3 | |
| HMS | | | 9.90 | 87–109 | 12.1–25.6 | |
| IAMC | QuEChERS | GC–MS/MS | 9.5 | 113–121 | 3.4–19.8 | This study |
| 4-MBC | | | 8.00 | 66–97 | 3.2–10.7 | |
| BP3 | | | 4.30 | 66–88 | 13.6–16.2 | |
| EDP | | | 12.1 | 103–106 | 2.3–15.4 | |
| EHMC | | | 8.00 | 107–123 | 3.4–22.6 | |

1H-BT: 1H-benzotriazole, 5Me-1H-BT: 5-methyl-1H-benzotriazole, TBHPBT: 2-(5-t-butyl-2-hydroxyphenyl) benzotriazole, 4-OH-HB: 4-hydroxy benzophenone. * LOQ values.

The developed method was applied to sludge samples collected from a domestic wastewater treatment plant. The concentrations of UV filters were found to be 66.9, 161.9, 54.5, <LOQ and 79.9 ng g$^{-1}$ for EHS, HMS, BP-3, EDP and EHMC, respectively. These results indicate that even in domestic wastewater sludge, high concentrations of UVFs could be found, and the sludge used for soil conditioning or fertilization purposes could raise environmental concerns.

## 4. Conclusions

In this study, eight different UVFs were extracted from sludge samples using a well-established QuEChERS methodology, and then, in-port derivatization was applied for more polar and less volatile analytes. The method combining QuEChERS and in-port derivatization, which ensures extra sensitivity, was shown to be accurate, reproducible and sensitive. This method also provided substantially reduced analysis time and solvent or chemical consumption compared to off-line derivatization. Method validation resulted in high recovery of UVFs (66–123%), meaning good accuracy; moreover, low inter-day RSD% (10.1–25.6) indicates high precision, and low values of LOD (<2.1 ng g$^{-1}$) and LOQ (<40 ng g$^{-1}$) show the sensitivity of the analysis for a matrix as complex as waste sludge. This study can be extended to the analysis of UVF degradation metabolites in sludge samples using QuEChERS with in-port derivatization followed by GC–MS/MS.

**Author Contributions:** C.Y. carried out the experiments, evaluation of results and preparation of the manuscript. F.N.E. and I.K. guided the experiments, edited the manuscript and evaluated the results. H.E. designed the experiments and guided the analysis. Research Assistant S.C.T. contributed to the analysis and experiments. All authors have read and agreed to the published version of the manuscript.

**Funding:** This research was funded by Dokuz Eylül University Scientific Research Projects (BAP) Coordination Unit] grant number [2021.KB.FEN.008].

**Institutional Review Board Statement:** Not applicable.

**Informed Consent Statement:** Not applicable.

**Data Availability Statement:** Not applicable.

**Acknowledgments:** We would like to thank Dokuz Eylül University Scientific Research Projects Coordination Unit (project number 2021.KB.FEN.008) for their financial support.

**Conflicts of Interest:** The authors declare no conflict of interest.

## References

1. Sabsevari, N.; Qiblawi, S.; Norton, S.; Fivenson, D. Sunscreens: UV filters to protect us: Part 1: Changing regulations and choices for optimal sun protection. *Int. J. Women's Dermatol.* **2021**, *7*, 28–44. [CrossRef] [PubMed]
2. Carve, M.; Allinson, G.; Nugegoda, D.; Shimeta, J. Trends in environmental and toxicity research on organic ultraviolet filters: A scientometric review. *Sci. Total Environ.* **2021**, *773*, 145628. [CrossRef] [PubMed]
3. Ramos, S.; Homem, V.; Alves, A.; Santos, L. Advances in analytical methods and occurrence of organic UV-filters in the environment—A review. *Sci. Total Environ.* **2015**, *526*, 278–311. [CrossRef] [PubMed]
4. Lambropoulou, D.A.; Giokas, D.L.; Sakkas, V.A.; Albanis, T.A.; Karayannis, M.I. Gas chromatographic determination of 2-hydroxy-4-methoxybenzophenone and octyldimethyl-p-aminobenzoic acid sunscreen agents in swimming pool and bathing waters by solid-phase microextraction. *J. Chromatography. A* **2002**, *967*, 243–253.5. [CrossRef] [PubMed]
5. Rodil, R.; Quintana, J.B.; López-Mahía, P.; Muniategui-Lorenzo, S.; Prada-Rodríguez, D. Multi-residue analytical method for the determination of emerging pollutants in water by solid-phase extraction and liquid chromatography–tandem mass spectrometry. *J. Chromatogr. A* **2009**, *1216*, 2958–2969. [CrossRef]
6. Jeon, K.; Chung, Y.; Ryu, J.-C. Simultaneous determination of benzophenone-type UV filters in water and soil by gas chromatography–mass spectrometry. *J. Chromatogr. A* **2006**, *1131*, 192–202. [CrossRef]
7. Votani, A.; Chisvert, A.; Giokas, D.L. On-line extraction coupled to liquid chromatographic analysis of hydrophobic organic compounds from complex solid samples—Application to the analysis of UV filters in soils and sediments. *J. Chromatogr. A* **2020**, *1610*, 460561. [CrossRef]
8. Li, M.; Sun, Q.; Li, Y.; Lv, M.; Lin, L.; Wu, Y.; Ashfaq, M.; Yu, C.-P. Simultaneous analysis of 45 pharmaceuticals and personalcare products in sludge by matrix solid-phase dispersion and liquid chromatography tandem mass spectrometry. *Anal. Bioanal. Chem.* **2016**, *408*, 4953–4964. [CrossRef]

9. Camino-Sanchez, F.J.; Zafra-Gomez, A.; Dorival-Garcia, N.; Juarez-Jimenez, B.; Vilchez, J.L. Determination of selected parabens, benzophenones, triclosan and triclocarban in agricultural soils after and before treatment with compost from sewage sludge: A lixiviation study. *Talanta* **2016**, *150*, 415–424. [CrossRef]
10. Sun, Q.; Li, M.; Ma, C.; Chen, X.; Xie, X.; Yu, C.-P. Seasonal and spatial variations of PPCP occurrence, removal and mass loading in three wastewater treatment plants located in different urbanization areas in Xiamen, China. *Environ. Pollut.* **2016**, *208*, 371–381. [CrossRef]
11. Schlumpf, M.; Cotton, B.; Conscience, M.; Haller, V.; Steinmann, B.; Lichtensteiger, W. In vitro and in vivo estrogenicity of UV screens. *Environ. Health Perspect.* **2001**, *109*, 239–244. [CrossRef]
12. Díaz-Cruz, M.S.; Barceló, D. Chemical analysis and ecotoxicological effects of organic UV-absorbing compounds in aquatic ecosystems. *Trends Anal. Chem.* **2009**, *28*, 708–716. [CrossRef]
13. Fivenson, D.; Sabsevari, N.; Qiblawi, S.; Blitz, J.; Norton, B.B.; Norton, S.A.; Fivenson, D. Sunscreens: UV filters to protect us: Part 2-Increasing awareness of UV filters and their potential toxicities to us and our environment. *Int. J. Women's Dermatol.* **2021**, *7*, 45–69. [CrossRef]
14. European Commission. DIRECTIVE 2008/105/EC OF THE EUROPEAN PARLIAMENT AND OF THE COUNCIL of 16 December 2008 on Environmental Quality Standards in the Field of Water Policy, Amending and Subsequently Repealing Council Directives 82/176/EEC, 83/513/EEC, 84/156/EEC, 84/491/EEC, 86/280/EEC and Amending Directive 2000/60/ EC of the European Parliament and of the Council. European Union: Brussels, Belgium, 2008. Available online: http://eur-lex.europa.eu/LexUriServ/LexUriServ.do?uri=OJ:L:2008:348:0084:0097:EN:PDF (accessed on 1 October 2022).
15. Directive, C. Council Directive 98/83/EC of 3 November 1998 on the Quality of Water Intended for Human Consumption. *Off. J. Eur. Communities* **1998**, 32–53. Available online: https://eur-lex.europa.eu/legal-content/EN/TXT/?uri=CELEX:31998L0083 (accessed on 13 September 2022).
16. Ramos, S.; Homem, V.; Santos, L. Development and optimization of a QuEChERS-GC–MS/MS methodology to analyse ultraviolet-filters and synthetic musks in sewage sludge. *Sci. Total Environ.* **2019**, *651*, 2606–2614. [CrossRef]
17. Chisvert, A.; Benedé, J.L.; Salvador, A. Current trends on the determination of organic UV filters in environmental water samples based on microextraction techniques-A review. *Anal. Chim. Acta* **2018**, *1034*, 22–38. [CrossRef]
18. Cadena-Aizaga, M.I.; Montesdeoca-Esponda, S.; Torres-Padrón, M.E.; Sosa-Ferrera, Z.; Santana-Rodríguez, J.J. Organic UV filters in marine environments: An update of analytical methodologies, occurrence and distribution. *Trends Environ. Anal. Chem.* **2020**, *25*, e00079. [CrossRef]
19. Zhang, Z.; Ren, R.; Li, Y.-F.; Kunisue, T.; Gao, D.; Kannan, K. Determination of benzotriazole and Benzophenone UV Filters in sediment and sewage sludge Environmental Science and Technology. *Environ. Sci. Technol.* **2011**, *45*, 3909–3916. [CrossRef]
20. Zhao, X.; Zhang, L.; Xu, L.; Liu, Y.; Song, W.W.; Zhu, F.J.; Li, Y.F.; Ma, W.L. Occurrence and fate of benzotriazoles UV filters in a typical residential wastewater treatment plant in Harbin, China. *Environ. Pollut.* **2017**, *227*, 215–222. [CrossRef]
21. Plagellat, C.; Kupper, T.; Furrer, R.; de Alencastro, L.F.; Grandjean, D.; Tarradellas, J. Concentrations and specific loads of UV filters in sewage sludge originating from a monitoring network in Switzerland. *Chemosphere* **2006**, *62*, 915–925. [CrossRef]
22. Kupper, T.; Plagellat, C.; Brändli, R.C.; de Alencastro, L.F.; Grandjean, D.; Tarradellas, J. Fate and removal of polycyclic musks, UV filters and biocides during wastewater treatment. *Water Res.* **2006**, *40*, 2603–2612. [CrossRef] [PubMed]
23. Negreira, N.; Rodríguez, I.; Rubí, E.; Cela, R. Optimization of pressurized liquid extraction and purification conditions for gas chromatography–mass spectrometry determination of UV filters in sludge. *J. Chromatogr. A* **2011**, *1218*, 211–217. [CrossRef] [PubMed]
24. Liu, Y.S.; Ying, G.-G.; Shareef, A.; Kookana, R.S. Simultaneous determination of benzotriazoles and ultraviolet filters in ground water, effluent and biosolid samples using gas chromatography–tandem mass spectrometry. *J. Chromatogr. A* **2011**, *1218*, 5328–5335. [CrossRef] [PubMed]
25. Liu, Y.S.; Ying, G.G.; Shareef, A.; Kookana, R.S. Occurrence and removal of benzotriazoles and ultraviolet filters in a municipal wastewater treatment plant. *Environ. Pollut.* **2012**, *165*, 225–232. [CrossRef] [PubMed]
26. Badia-Fabregat, M.; Rodriguez-Rodriguez, C.E.; Gago-Ferrero, P.; Olivares, A.; Pina, B.; Diaz-Cruz, M.S.; Vicent, T.; Barcelo, D.; Caminal, G. Degradation of UV filters in sewage sludge and 4-MBC in liquid medium by the ligninolytic fungus Trametes versicolor. *J. Environ. Manag.* **2012**, *104*, 114–120. [CrossRef]
27. Gago-Ferrero, P.; Diaz-Cruz, M.S.; Barcelo, D. Occurrence of multiclass UV filters in treated sewage sludge from wastewater treatment plants. *Chemosphere* **2011**, *84*, 1158–1165. [CrossRef]
28. Rodil, R.; Schrader, S.; Moeder, M. Pressurised membrane-assisted liquid extraction of UV filters from sludge. *J. Chromatogr. A* **2009**, *1216*, 8851–8858. [CrossRef]
29. Rodríguez-Rodrigueza, C.E.; Barónc, E.; Gago-Ferreroc, P.; Jeli, A.; Llorcac, M.; Farréc, M.; Díaz-Cruz, M.S.; Eljarrat, E.; Petrović, M.; Caminala, G.; et al. Removal of pharmaceuticals, polybrominated flame retardants and UV-filters from sludge by the fungus Trametes versicolor in bioslurry reactor. *J. Hazard. Mater.* **2012**, *233*, 235–243. [CrossRef]
30. Langford, K.H.; Reid, M.J.; Fjeld, E.; Øxnevad, S.; Thomas, K.V. Environmental occurrence and risk of organic UV filters and stabilizers in multiple matrices in Norway. *Environ. Int.* **2015**, *80*, 1–7. [CrossRef]
31. Vela-Soria, F.; Iribarne-Durán, L.M.; Mustieles, V.; Jiménez-Díaz, I.; Fernández, M.F.; Olea, N. QuEChERS and ultra-high performance liquid chromatography–tandem mass spectrometry method for the determination of parabens and ultraviolet filters in human milk samples. *J. Chromatogr. A* **2018**, *1546*, 1–9. [CrossRef]

32. Picot Groz, M.; Martinez Bueno, M.J.; Rosain, D.; Fenet, H.; Casellas, C.; Pereira, C.; Maria, V.; Bebianno, M.J.; Gomez, E. Detection of emerging contaminants (UV filters, UV stabilizers and musks) in marine mussels from Portuguese coast by QuEChERS extraction and GC-MS/MS. *Sci. Total Environ.* **2014**, *493*, 162–169. [CrossRef]
33. Cerqueira, M.B.R.; Guilherme, J.R.; Caldas, S.S.; Martins, M.L.; Zanella, R.Z.; Primel, E.G. Evaluation of the QuEChERS method for the extraction of pharmaceuticals and personal care products from drinking-water treatment sludge with determination by UPLC-ESI-MS/MS. *Chemosphere* **2014**, *107*, 74–82. [CrossRef]
34. Herrero, P.; Borrull, F.; Pocurull, E.; Marcé, R.M. A quick, easy, cheap, effective, rugged and safe extraction method followed by liquid chromatography-(Orbitrap) high resolution mass spectrometry to determine benzotriazole, benzothiazole and benzenesulfonamide derivates in sewage sludge. *J. Chromatogr. A* **2014**, *1339*, 34–41. [CrossRef]
35. Gago-Ferreroa, P.; Mastroiannia, N.; Díaz-Cruz, M.S.; Barceló, D. Fully automated determination of nine ultraviolet filters and transformation products in natural waters and wastewaters by on-line solid phase extraction–liquid chromatography–tandem mass spectrometry. *J. Chromatogr. A* **2013**, *1294*, 106–116. [CrossRef]
36. Wick, A.; Fink, G.; Ternes, T.A. Comparison of electrospray ionization and atmospheric pressure chemical ionization for multi-residue analysis of biocides, UV-filters and benzothiazoles in aqueous matrices and activated sludge by liquid chromatography–tandem mass spectrometry. *J. Chromatogr. A* **2010**, *1217*, 2088–2103. [CrossRef]
37. Yücel, C.; Ertaş, H.; Ertaş, F.N.; Karapinar, İ. Determination of UV Filters in Surface Water by VA-DLLME-SFOD Technique Coupled with GC–MS/MS. *Clean–Soil Air Water* **2022**, *50*, 2100246. [CrossRef]
38. Peters, F.T.; Drummer, O.H.; Musshoff, F. Validation of new methods. *Forensic Sci. Int.* **2007**, *165*, 216–224. [CrossRef]
39. EURACHEM / CITAC Guide CG 4 Quantifying Uncertainty in Analytical Measurement. 2012. Available online: https://www.eurachem.org/images/stories/Guides/pdf/QUAM2012_P1.pdf (accessed on 10 August 2022).
40. Simões, N.G.; Cardoso, V.V.; Ferreira, E.; Benoliel, M.J.; Almeida, C.M.M. Experimental and statistical validation of SPME-GC–MS analysis of phenol and chlorophenols in raw and treated water. *Chemosphere* **2007**, *68*, 501–510. [CrossRef]
41. Eurachem Guide. *The Fitness for Purpose of Analytical Methods—A Laboratory Guide to Method Validation and Related Topics*; Eurachem Guide: Zug, Switzerland, 2014.
42. Shah, V.P.; Midha, K.K.; Dighe, S.; McGilveray, I.J.; Skelly, J.P.; Yacobi, A.; Layloff, T.; Viswanathan, C.T.; Cook, C.E.; McDowall, R.D.; et al. Analytical methods validation: Bioavailability, bioequivalence and pharmacokinetic studies. *J. Pharm. Sci.* **1992**, *81*, 588–592.
43. Shah, V.P.; Midha, K.K.; Findlay, J.W.; Hill, H.M.; Hulse, J.D.; McGilveray, I.J.; McKay, G.; Miller, K.J.; Patnaik, R.N.; Powell, M.L.; et al. Bioanalytical method validation—A revisit with a decade of progress. *Pharm. Res.* **2000**, *17*, 1551–1557. [CrossRef]
44. Food and Drug Administration, Q2B Validation of Analytical Procedures: Methodology. Available online: https://www.fda.gov/regulatory-information/search-fda-guidance-documents/q2b-validation-analytical-procedures-methodology (accessed on 25 September 2022).
45. European Medicines Agency. *ICH Topic Q 2 (R1) Validation of Analytical Procedures: Text and Methodology*; CPMP/ICH/381/95; European Medicines Agency: London, UK, 1995.
46. Pan, L.; Pawliszyn, J. Derivatization/Solid-Phase Microextraction: New, Approach to Polar Analytes. *Anal. Chem.* **1997**, *69*, 196–205. [CrossRef]
47. Basheer, C.; Lee, K. Analysis of endocrine disrupting alkylphenols, chlorophenols and bisphenol-A using hollow fiber-protected liquid-phase microextraction coupled with injection port-derivatization gas chromatography–mass spectrometry. *J. Chromatogr. A* **2004**, *1057*, 163–169. [CrossRef] [PubMed]
48. Bizkarguenaga, E.; Iparragirre, A.; Navarro, P.; Olivares, M.; Prieto, A.; Vallejo, A.; Zuloaga, O. In-port derivatization after sorptive extractions-Review. *J. Chromatogr. A* **2013**, *1296*, 36–46. [CrossRef] [PubMed]

Article

# On-Site Multisample Determination of Chlorogenic Acid in Green Coffee by Chemiluminiscent Imaging

Sergi Mallorca-Cebria, Yolanda Moliner-Martinez *, Carmen Molins-Legua and Pilar Campins-Falcó

MINTOTA Research Group, Departament de Química Analítica, Facultad de Química, Universitat de Valencia. C/Doctor Moliner 50, 46100 Burjassot, Spain
* Correspondence: yolanda.moliner@uv.es

**Abstract:** The potential of antioxidants in preventing several diseases has attracted great attention in recent years. Indeed, these products are part of a multi-billion industry. However, there is a lack of scientific information about safety, quality, doses, and changes over time. In the present work, a simple multisample methodology based on chemiluminiscent imaging to determine chlorogenic acid (CHLA) in green coffee samples has been proposed. The multi-chemiluminiscent response was obtained after a luminol-persulfate reaction at pH 10.8 in a multiplate followed by image capture with a charge-coupled device (CCD) camera as a readout system. The chemiluminiscent image was used as an analytical response by measuring the luminescent intensity at 0 °C with the CCD camera. Under the optimal conditions, the detection limit was 20 µM and precision was also adequate with RSD < 12%. The accuracy of the proposed system was evaluated by studying the matrix effect, using a standard addition method. Recoveries of chlorogenic acid ranged from 93–94%. The use of the CCD camera demonstrated advantages such as analysis by image inspection, portability, and easy-handling which is of particular relevance in the application for quality control in industries. Furthermore, multisample analysis was allowed by one single image saving time, energy, and cost. The proposed methodology is a promising sustainable analytical tool for quality control to ensure green coffee safety through dosage control and proper labelling preventing potential frauds.

**Keywords:** antioxidants; chlorogenic acid; chemiluminiscence; CCD camera; green coffee; on-site

Citation: Mallorca-Cebria, S.; Moliner-Martinez, Y.; Molins-Legua, C.; Campins-Falcó, P. On-Site Multisample Determination of Chlorogenic Acid in Green Coffee by Chemiluminiscent Imaging. *Methods Protoc.* **2023**, *6*, 20. https://doi.org/10.3390/mps6010020

Academic Editors: Victoria Samanidou, Verónica Pino and Natasa Kalogiouri

Received: 16 December 2022
Revised: 6 February 2023
Accepted: 10 February 2023
Published: 14 February 2023

**Copyright:** © 2023 by the authors. Licensee MDPI, Basel, Switzerland. This article is an open access article distributed under the terms and conditions of the Creative Commons Attribution (CC BY) license (https://creativecommons.org/licenses/by/4.0/).

## 1. Introduction

The increasing consumption of antioxidants raises public health concerns about their efficiency, safety, and real benefits for human health. Antioxidants are widely available and commonly used, especially in the women's health market [1]. However, there is a lack of information about the composition, doses, potential degradation, and function of antioxidants [2]. Mainly, the benefits of antioxidant supplements are associated with their interaction with free radicals and hence protect health against oxidative stress [3]. The interaction mechanism depends on the chemical nature and reactivity. These compounds are classified as direct and indirect antioxidants and they mediate oxidative stress through different mechanisms. Nevertheless, there are controversial results and conclusions about the real level and function of these compounds [4].

What is certain, is that the antioxidant nutraceutical market is now a multi-billion dollar industry experiencing a continuous increase, and this has led to regulations that ensure safety, proper labelling, and more information about the ingredients, which means there is a real need to ensure the concentration of the antioxidant compounds, and that they are within safety limits, in particular in quality control studies [5].

Focusing on direct antioxidants, chlorogenic acid (CHLA) is an abundant water-soluble antioxidant in the human diet. Indeed, it is available from a wide variety food and beverage sources. In particular, CHLA is linked to green coffee and green coffee extract-based products' health benefits. That fact, together with the relevance of the global coffee market, has led to a growing interest in the development of analytical methods for the quantification of this antioxidant [6].

In this context, the development of analytical strategies that facilitates the determination of CHLA, in commercial products is mandatory to help the regulation, labelling, and warming of potential side effects [7]. Basically, the main methods to determine this compound are based on separation techniques such as HPLC and UPLC, coupled to different detectors such as DAD, luminescent, and MS [8–12]. Their performance has been demonstrated, yet they required specialized laboratory instrumentation, trained personnel, and a large instrumental investment. Thus, more recently simple, cost-effective, rapid, easy-handling, and portable analytical procedures have been proposed.

These new strategies are principally based on optical readouts with the added value of providing visual responses that allow semiquantitative determination by visual or image inspection and on-site determination [13]. Colorimetric-based sensors using paper-based devices and a metal nanooxide array have been recently published [14,15].

Among the different on-site strategies, chemiluminiscence (CL) is attracting attention since it allows for the achievement of sensitivity in short analysis times [16]. Going a step forward, chemiluminiscent imaging is a powerful alternative since it combines the typical CL properties with visual inspection and portability [12,17]. In this case, the use of charge-couple device (CCD) cameras has been demonstrated to be a feasible detector for this aim [18]. Fundamentally, the photons produced in the hemiluminescent reaction are captured by the CCD camera and subsequently, an image is displayed based on the light generated, the intensity measurement can be correlated with the target analyte. The key parameters in the application of these systems are mainly based on the control of experimental variables since the CL efficiency depends on the kinetic emission CCD camera response time. From the analytical point of view, there is scarce knowledge of the potential application of CCD cameras as analytical response read-out systems. This strategy has been proposed for environmental analysis [18], however, other application areas and in particular for quality control of human consumption products, are still unexplored.

The objective of the present work was to demonstrate the utility of chemiluminiscent imaging employing the CCD camera to estimate CHLA in natural products, in particular green coffee samples. The novelty relies on the combination of an on-site portable device with high-performance analytical parameters for quality control of commercial products. Optimization of the reaction parameters was aided by the use of a portable luminometer. For this aim, the instrumental parameters were optimized and the analytical properties were established. Finally, the practical application and validation of the proposed methodology were evaluated by analysing commercial samples.

## 2. Experimental Section

### 2.1. Reagents

Sodium hypochlorite 10% was obtained from Panreac (Barcelona, Spain). Sodium persulfate, Trolox, CHLA, and luminol were purchased from Sigma-Aldrich (Taufkirchen, Germany). Sodium carbonate and sodium hydroxide were from VWR Chemicals (Radnor, PA, USA).

Standard solutions of sodium persulfate (50 mM) in carbonate buffer 0.3 M (pH 10.8) were prepared weekly. Trolox (1.1 mM), luminol (20 mM), and CHLA (2.0 mM) were prepared in carbonate buffer (0.3 M, pH 10.8). Working standard solutions of Trolox and CHLA were prepared by adequate dilution in carbonate buffer.

### 2.2. Equipment

The CL response was measured by using a 338L Mono CCD Camera themoelectrically cooled at 5 °C (Atik Cameras, Norwich, UK). In addition, an LB 9509 portable tube luminometer (Berthold Technologies, Bad Wildbad, Baden-Württemberg, Germany) was also used to determine the optima chemiluminiscent reaction conditions.

## 2.3. Procedure

### 2.3.1. Optimization and Measurement of the CL Response

In the preliminary studies, the optimization of the reaction parameters, such as oxidant, concentration, pH, and analysis time was carried out using the luminometer and CHLA analytical grade standards (see Section 2.1). Oxidants such as sodium persulfate and sodium hypochlorite were evaluated. For this aim, 100 µL of luminol was mixed with 100 µL of persulfate or hypochlorite of luminol and the response was registered every 5 s, for 1 min. The study of persulfate and luminol concentrations were performed in the range of 0.7–10 mM, and 1.4–4.8 mM, respectively, by measuring the signal at 60 s. Finally, the influence of pH was studied in the interval 10.0–11.5, using persulfate (10 mM) and luminol (4.8 mm).

Once the optimal reaction conditions were established, the CCD camera instrumental parameters were studied to monitor the CL imaging response.

### 2.3.2. CCD Measurements

The CCD camera was studied as a readout system. In this case, contrast, temperature, and exposition time were previously optimized. Under the different conditions, the images were processed by the software Artemis Capture, and the luminescent intensity (I) for each spot was registered. Standards and samples were deposited on a multiplate and placed inside the CCD camera dark chamber for its measurement. The device used to perform the CCD measurements is shown in Figure 1. It should be noted that exposition distance was optimized in previous work [18].

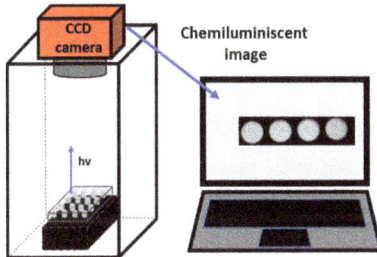

**Figure 1.** Schematic diagram of the CCD camera dark chamber used for the determination of CHLA.

The measurement was performed in a multiplate by mixing luminol (100 µL) and persulfate (100 µL). After 5 min, 100 µL of the blank, working standards, or samples, prepared in a carbonate buffer (0.3 M, pH 10.8), were added, and the image was registered. The working concentration interval was up to 500 µM for CHLA. Trolox was also measured as another example of the antioxidant analyte.

## 2.4. Analysis of Commercial Green Coffee Samples

The application of the proposed method was validated by the analysis of three commercial green coffee samples acquired in a local store. Sample preparation was carried out as follows: a capsule of each sample was dissolved in ultrapure water and sonicated during 5 min, to extract CHLA. Subsequently, the extract was diluted in carbonate buffer (0.3 M, pH 10.8). Samples were measured following the proposed procedure. The analysis was performed in triplicate. In addition, the accuracy of the proposed method was evaluated by using the standard addition method (SAM). For this aim, the calibration curve was performed following the procedure described in Section 2.3 and spiking the sample with known concentrations of CHLA standards.

## 3. Procedure

### 3.1. Study of Chemiluminiscent Imaging as an Analytical Response

In the previous experiments, the CL response was optimised considering the hypothesis that the reaction between luminol and an oxidant in basic media, induces the generation of a luminescent signal that can be monitored, and the presence of antioxidants such as chlorogenic acid may inhibit the CL signal as a function of the concentration. Hence, the first step was the optimization of the CL reaction, taking into account that the CCD camera will be the readout system, which means that detection time is a key parameter. Figure 2 shows the variation of the CL response as a function of the experimental parameters (measurements done with the portable luminometer).

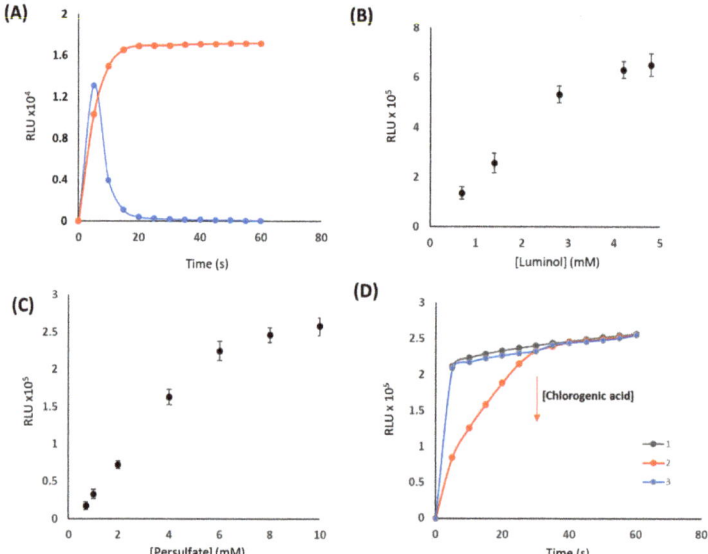

**Figure 2.** (**A**) Variation of the luminescent signal (RLU) produced by luminol using 0.25 mM hypochlorite (blue) and 0.7 mM persulfate (red); Luminol = 1.4 mM, carbonate buffer 0.3 M, pH 10.8. (**B**) Variation of the RLU as a function of luminol concentration, (**C**) Variation of the RLU as a function of persulfate concentration, (**D**) Variation of the RLU signal in presence of CHLA (50 µM, pH 10.8) for blank (1, grey), luminol-persulfate, CHLA addition at t = 5 min (2, red) and persulfate-CHLA luminol addition at t = 10 min (3, blue). Conditions: 10 mM persulfate, 1.4 mM luminol.

As can be seen in Figure 2A, the use of hypochlorite as an oxidant resulted in a fast reaction kinetic with a signal decay in 10 s. Meanwhile, persulfate provided a constant CL signal up to 60 s. Therefore, persulfate was selected as an oxidant, taking into account the exposition time that subsequently will be needed for the measurement in the CCD camera. Regarding luminol and persulfate concentrations (see Figure 2B,C), 4.8 mM, and 10 mM were the optimum concentration, respectively, since they maximized the CL signal, and hence the sensitivity of the target analytes by inhibition of this CL response. It is well known the dependence of CL reactions with pH, thus, this parameter was also studied. NaOH and carbonate buffer were studied. The results indicated that carbonate buffer 0.3 M and pH 10.8, provided satisfactory intensity, taking into account the concretion level of the samples of interest.

Under the above-mentioned conditions, the influence of the addition of antioxidants was evaluated. Previous reports indicated that antioxidants give rise to a decrease in the CL response and the mechanism involved is not yet clear, since some authors pointed to CL quenching in addition to the inhibition of luminol oxidation by the antioxidants that has also been reported [17]. In this scenario, different strategies were evaluated. In the first case, persulfate and CHLA were mixed, and after 10 min, luminol was added. In the second experiment, luminol and persulfate were mixed, and after 5 min, CHLA was added. Figure 1D shows the results obtained. As can be seen, in the second condition, a decrease of the CL signal was observed with the addition of CHLA. Hence, that was the strategy selected for further experiments.

By another hand, the variation of the CL decrease as a function of CHLA concentration was also studied in order to establish the potential use as a quantitative methodology. Firstly, the reaction time between persulfate and luminol, before the addition of the antioxidant was studied. For this aim, additional times of 180 and 300 s were studied. Figure 3A shows the variation of the response as a function of time, and at two antioxidant addition times. The addition of the target analytes at 300 s was advantageous since the stability of the response was higher than at 180 s, more likely due to a slower kinetic of the antioxidants since, persulfate in the reaction media is lower at 300 s than at 180 s. This stability was a key parameter, since, the CCD camera required a stability interval to properly register the image. The inclusion criteria, in this case, was that the signal was stable for up to 30 s. Figure 3B shows the variation of the kinetics at different concentration levels. By measuring the signal at 30 s after the addition of chlorogenic acid, the CL decrease could be linearly correlated with the concentration level (sensitivity = $-1400 \pm 50 \ \mu M^{-1}$, $R^2$ = 0.9930), hence the quantitative potential was demonstrated under these conditions with a detection limit, LOD = 2 µM.

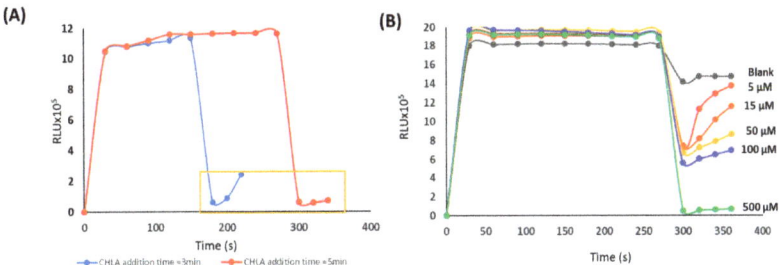

**Figure 3.** (**A**) Variation of the RLU in presence of CHLA (500 µM) after 180 s (blue) and 300 s (red). (**B**) Variation of the RLU over CHLA concentration under the optima reaction conditions.

Remarkably, it should be considered that in presence of other antioxidant compounds, the decrease in the luminescent signal would be a result of the total antioxidant content extracted under the extraction conditions described in Section 2.4. In order to demonstrate this, the RLU signal for trolox was also measured. The results indicated the same behaviour as CHLA, however, the decrease of the luminescent signal was slightly higher, achieving a detection limit of 0.7 µM. Hence, in this case, the luminescent signal would provide a measurement of the total content of CL active compounds under the extraction conditions.

Once the reaction parameters were optimized, the CCD camera instrumental parameters, the contrast parameters, exposition time, and temperature were studied. The CL response was established by capturing the image and processing that image with Artemis Capture software. Figure 4 shows the schematic representation of the set-up and the images obtained for different blanks measured at different times under the contrast parameters white 2000, black 450, and log 1.5 for CHLA. It should be noted that the image shown in Figure 4 was a real image, where the responses were specific dots for each blank. In this case, the analytical response was the luminescent intensity (I) of each sample, and hence the balance between black and white intensity and grey contrast was fundamental to obtain adequate image quality. As can be seen, the profile of the CL image signal was similar to that measured with the luminometer. Hence, it was expected, that the addition of CHLA to the plate would result in a decrease in the CL that can be monitored by the CCD camera.

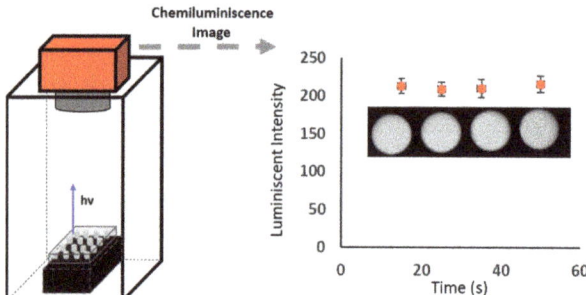

**Figure 4.** Schematic set-up of the CCD camera to measure CL image, the real image of the variation of the bank signal as a function of time, and the representation of the luminescent intensity for each image.

To demonstrate the response as a function of CHLA concentration, the target analyte was added and the experiment was performed at two temperatures. The results indicated that the temperature of the CCD camera was a key parameter. Mainly at high a temperature, the background noise or dark current gave rise to a low-resolution image. Meanwhile, working at 0 °C, the resolution was satisfactory, and the difference in the blank signal compared with the CHLA standard could be monitored by the image. Figure 5A shows the CL images carried out at 0 and 23 °C for CHLA compared with blanks at the same temperature.

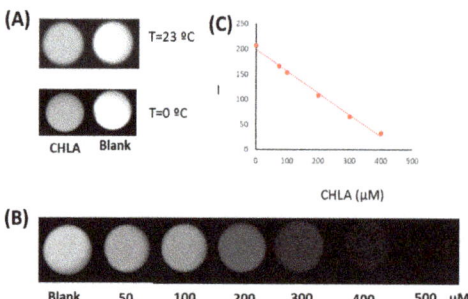

**Figure 5.** (**A**) Effect of the CCD camera temperature on the chemiluminiscent image (CHLA 100 µM), (**B**) Variation of the chemiluminiscent intensity over CHLA concentration, and (**C**) Correlation of the CL intensity as a function of CHLA concentration.

The variation of the luminescent intensity with the exposition time can be seen in Figure 3, the readout measurement is stable up to 30 s due to the chemistry of the reaction. Hence, exposition times from 5 to 20 s were studied. The results indicated, that at times lower than 10 s, the image resolution was poor. At exposition times higher than 10 s, the resolution was satisfactory, however, there was no improvement of the longer times (20 s) used. Therefore, 10 s was selected for further experiments.

Finally, and under the above-mentioned instrumental parameters, the variation of the CCD camera response was studied as a function of the concentration of CHLA. Figure 5B shows the image obtained. These results clearly demonstrated that the decrease of the CL signal as a function of the concentration of CHLA was linearly correlated (Figure 5C). It should be noted that the image parameter black and white, which enhanced the intensity of black and white, and the grey contrast up to 1.5 (log parameter) gave rise to an adequate image quality that allowed the determination of the target analyte up to 400 µM.

### 3.2. Analytical Parameters of the Procedure

Table 1 summarizes the analytical parameters such as sensitivity as the slope of the calibration curve, linearity, and detection limit. Moreover, intra-and interday precision was also evaluated as the relative standard deviation (%RSD)

**Table 1.** Analytical parameters and precision (%RSD) values for the determination of CHLA by the CCD camera.

|  | Analytical Parameters |
|---|---|
| Working interval ($\mu$M) | 50–400 |
| $a \pm s_a$ | $200 \pm 4$ |
| $b \pm s_b$ ($\mu M^{-1}$) | $0.470 \pm 0.017$ |
| $R^2$ | 0.999 |
| LOD ($\mu$M) | 20 |
| RSD $_{intraday}$ (%) | 2.0 |
| RSD $_{interday}$ (%) | 12.0 |

a: intercept; $s_a$: standard deviation of a; b: slope, $s_b$: standard deviation of b; LOD: detection limit, RSD: relative standard deviation.

The results indicated that when using a chemiluminiscent image as an analytical response, the working interval was up to 400 $\mu$M, and the limit of detection was 20 $\mu$M. This LOD was satisfactory for the application in dietary products, however, a more sensitive determination will require the use of other analytical responses, such as the use of a portable luminometer. Intraday and interday precision was also evaluated. In both cases, satisfactory RSD values were achieved for practical applications.

As mentioned before, previous methods based on HPLC have been proposed, and the comparison reveals that, in general, chromatographic methods provided LOD slightly lower than the LOD obtained with the CCD method, recoveries were also satisfactory, and, obviously, selectivity was better when using separation methods [19]. Capillary electrophoresis with CL detection has resulted in an adequate method for the analysis of natural products, however, the LOD was higher than that obtained with the proposed method [20]. NIR has also been proposed for CHLA determination, with successful results. However, multivariate calibration was necessary [21]. Regarding the in situ-based analytical systems, sensors and biosensors have also been studied in depth. A review focused on this topic has been recently published by Munteanu and Apetrei [6]. Recent advances in sensitive materials for electrochemical sensors and biosensors have led to methodologies, with low $\mu$M detection limits, which is an advantage if the content of this type of antioxidant is at a trace level [22]. However, precision depends on the accuracy and expertise in the preparation of the sensitive material. Therefore, the proposed methodology shows some advantages over these previously proposed methods; since the determination of CHLA can be performed by image inspection, a semiqualitative analysis for sample screening is possible. In addition, quantitative analysis can be also carried out for samples with contents higher than 20 $\mu$M. By another hand, cost of the CCD is lower than chromatographic electrophoretic methods.

### 3.3. Procedure Validation: Application to the Analysis of Green Coffee Samples

In order to validate the proposed procedure for its practical application, the analysis of three different green coffee samples, being CHLA the active compound. In addition, and in order to evaluate the matrix effect and the potential interference of other ingredients present in the sample composition, the standard addition method (SAM) and a recovery study were carried out and the slopes of the calibration curves were compared with external calibration. Figure 6 shows the images of both calibration strategies. As was expected, the luminosity intensity of the blank in the external calibration was higher than in the first sample of the standard addition method due to the presence of CHLA, and in both cases, there was a decrease in the CL intensity over the CHLA concentration.

The slopes of the SAM calibration were $-0.49 \pm 0.3$, $-0.37 \pm 0.02$, $-0.37 \pm 0.01$ $\mu M^{-1}$ for samples 1, 2, and 3, respectively. The statistical analysis of the SAM slope and the slope of the external calibration (see Table 1) demonstrated that the sample matrix may influence the luminescent intensity, and hence, it may be necessary for SAM as a calibration method. In the samples analysed in this work, only sample 1 was not affected by the matrix effect, since there were no statistical differences in the slope of the calibration graph. However, that influence depended on the commercial product, the matrix effect was observed in samples 2 and 3. The calculated CHLA contents were $145 \pm 10$, $220 \pm 12$, and 204 mg CHLA/capsule for sample 1, sample 2, and sample 3, respectively. These results were then compared with the labels of the commercial products. Sample 1 and sample 2 were labelled with a CHLA content of 157 mg/capsule and 234 mg/capsule, respectively. The statistical comparison with the quantitative analysis carried out with the proposed methodology revealed no statistical differences, and hence, accuracy was demonstrated. In the case of sample 3, the content of CHLA was not labelled. However, the analysis of the sample determined 207 mg/capsule, expressed as CHLA. Recovery values were in the range between $93 \pm 8$ and $94 \pm 9$ %. These results indicated that the proposed methodology can be a relevant methodology to evaluate the content of CHLA in the analysed samples, since commercially, some of these products are not properly labelled, and that matrix effect has to be evaluated as a function of the analysed samples. It has been demonstrated that standard addition methods provided satisfactory results in the case of the matrix effect.

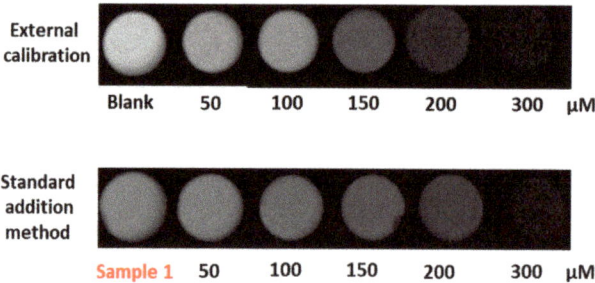

**Figure 6.** Comparison of the CL image for external calibration and standard addition method in sample 1.

## 4. Conclusions

In this work, an on-site methodology based on chemiluminscent imaging has been proposed to estimate the concentration of CHLA in green coffee samples. The fundamentals of the proposed strategy relied on the use of the CL to decrease the presence of antioxidants, in particular CHLA, produced in the reaction luminol-persulfate under basic conditions. Under the optima chemical conditions, the results indicated that the addition of CHLA induced a stable analytical response that can be used as an analytical signal for the CCD camera as the detection system. The image taken by the CCD camera was processed. At a T = 0 °C and an exposition time of 10 s, the decrease in the luminescent intensity was linearly correlated with CHLA concentration, with a LOD of 20 $\mu M$ and RSD values up to 12%. The satisfactory analytical performance was validated by applying the proposed strategy for the analysis of commercial green coffee samples. The results demonstrated that the matrix effect should be considered when analysing real samples. SAM was adequate for the calibration since the sample matrix may influence the luminosity intensity, however, that effect depended on the sample. The analysis of samples was validated by comparing the results with the labelled products, indicating that satisfactory accuracy was obtained. Hence, it can be concluded that the proposed strategy can be a potential on-site tool to estimate the concentration of antioxidants in dietary products. The analysis of the chemiluminscent image in the set-up allowed the multisample detection resulting in a cost-

and energy-efficient protocol with reagent consumption at μM level. That demonstrated that the proposed method is a promising sustainable analytical tool for quality control to ensure supplement safety and proper labelling, preventing potential fraud.

**Author Contributions:** S.M.-C.: investigation, methodology, validation. Y.M.-M.: conceptualization, funding acquisition, supervision, revision. C.M.-L.: conceptualization, supervision. P.C.-F.: conceptualization, project administration, funding acquisition. All authors have read and agreed to the published version of the manuscript.

**Funding:** This research has been funded by the EU and the Gobierno de España MCI-AEI (PID2021-124554NB-I00), the Generalitat Valenciana (PROMETEO Program 2020/078 and EU FEDER- Generalitat Valenciana (ID-FEDER/2018/049).

**Data Availability Statement:** Data is contained within the article.

**Acknowledgments:** Authors acknowledge the EU and the Gobierno de España MCI-AEI (PID2021-124554NB-I00) and the Generalitat Valenciana (PROMETEO Program 2020/078 and EU FEDER-Generalitat Valenciana (ID-FEDER/2018/049).

**Conflicts of Interest:** The authors declare no conflict of interest.

# References

1. Rogge, M.; Kumar, P.; Grundmann, O.; ACCP Public Policy Committee. Front-Line Health Care Professionals Lack Critical Knowledge in Dietary Supplement and Nutraceutical Products: A Call to Action for Comprehensive Educational Opportunities. *J. Clin. Pharmacol.* **2022**, *62*, 17–19. [CrossRef] [PubMed]
2. Binns, C.W.; Lee, M.K.; Lee, A.H. Problems and Prospects: Public health regulations of dietary supplements. *Annu. Rev. Public Health* **2018**, *39*, 403–420. [CrossRef] [PubMed]
3. Joko, S.; Watanabe, M.; Fuda, H.; Takeda, S.; Furukawa, T.; Hui, S.-P.; Shrestha, R.; Chiba, H. Comparison of chemical structures and cytoprotection abilities between direct and indirect antioxidants. *J. Funct. Foods* **2017**, *35*, 245. [CrossRef]
4. Bandara, S.B.; Urban, A.; Liang, L.G.; Parker, J.; Fung, E.; Maier, A. Active pharmaceutical contaminants in dietary supplements: A tier-based risk assessment approach. *Regul. Toxicol. Pharmacol.* **2022**, *123*, 104955. [CrossRef]
5. Coskun, S.H.; Wise, S.A.; Kuszak, A.J. The Importance of Reference Materials and Method Validation for Advancing Research on the Health Effects of Dietary Supplements and Other Natural Products. *Front. Nutr.* **2021**, *8*, 786261. [CrossRef]
6. Munteanu, I.G.; Apetrei, C. A review on electrochemical sensors and biosensors used for chlorogenic acid electroanalysis. *Int. J. Mol. Sci.* **2021**, *22*, 13138. [CrossRef]
7. Vidal-Casanella, O.; Núñez, O.; Granados, M.; Saurina, J.; Sentellas, S. Sentellas. Analytical methods for exploring nutraceuticals based on phenolic acids and polyphenols. *Appl. Sci.* **2021**, *11*, 8276. [CrossRef]
8. Jia, W.; Shi, L.; Zhang, F.; Chang, J.; Chu, X.G. High-Throughput mass spectrometry scheme for screening and quantification of flavonoids in antioxidants nutraceuticals. *J. Chromatogr. A* **2019**, *1608*, 460408. [CrossRef]
9. Fantoukh, O.I.; Wang, Y.H.; Parveen, A.; Hawwal, M.F.; Ali, Z.; Al-Hamoud, G.A.; Chittiboyina, A.G.; Joubert, E.; Viljoen, A.; Khan, I.A. Chemical fingerprinting profile and targeted quantitative analysis of phenolic compounds from Rooibos tea and dietary supplements by using UHPLC-PDA-MS. *Separations* **2022**, *9*, 159. [CrossRef]
10. Brzezicha, J.; Błażejewicz, D.; Brzezińska, J.; Grembecka, M. Green coffee vs dietary supplements: A comparative analysis of bioactive compounds and antioxidant activity. *Food Chem. Toxicol.* **2021**, *155*, 112377. [CrossRef]
11. Saar-Reismaa, P.; Koel, M.; Tarto, R.; Vaher, M. Extraction of bioactive compounds from Dpsacus fullonum leaves using Deep eutectic solvents. *J. Chromatogr. A* **2022**, *1677*, 463330. [CrossRef] [PubMed]
12. Krzyminski, K.K.; Roshal, A.D.; Rudnicki-Velasquez, P.B.; Zamojc, K. On the use of acridinium indicators for the chemiluminiscent determination of the total antioxidant capacity of dietary supplements. *Luminescence* **2019**, *34*, 512. [CrossRef] [PubMed]
13. Jornet-Martínez, N.; Moliner-Martínez, Y.; Molins-Legua, C.; Campíns-Falcó, P. Trends for the Development of In Situ Analysis Devices. In *Encyclopedia of Analytical Chemistry*; John Wiley & Sons, Ltd.: Hoboken, NJ, USA, 2017; pp. 1–23.
14. Poppa, C.V.; Visilescu, A.; Litescu, S.C.; Albu, C.; Danet, A.F. Metal nano-oxide based colorimetric sensor array for the determination of polyphenols with antioxidant properties. *Anal. Lett.* **2020**, *53*, 627. [CrossRef]
15. Puangbanlang, C.; Sirivibulkovit, K.; Nacaprichah, D.; Sameenoi, Y. A paper-based device for simultaneous determination of antioxidant activity and total phenolic content in food samples. *Talanta* **2019**, *198*, 542. [CrossRef] [PubMed]
16. Tripathi, P.; Kumar, A.; Sachan, M.; Gupta, S.; Nara, S. Aptamer-gold nanozyme based competitive lateral flow assay for rapid detection of CA125 in human serum. *Biosens. Bioelectr.* **2020**, *165*, 112368. [CrossRef]
17. Chládková, G.; Kunovská, K.; Chocholouš, P.; Polášek, M.; Sklenářová, H. Automatic Screening of Antioxidants Based on the Evaluation of Kinetics of Suppression of Chemiluminescence in a Luminol-Hydrogen Peroxide System Using a Sequential Injection Analysis Setup with a Flow-Batch Detection Cell. *Anal. Meth.* **2019**, *11*, 2531. [CrossRef]

18. González-Fuenzalida, R.A.; Molins-Legua, C.; Calabria, D.; Mirasoli, M.; Guardigli, M.; Roda, A.; Campins-Falcó, P. Sustainable and green persulfate-based chemiluminescent method for on-site estimation of chemical oxygen demand in water. *Anal. Chim. Acta* **2022**, *1223*, 340196. [CrossRef]
19. Velkiska-Markovska, L.; Jankulovska, M.S.; Petanovska-Ilievska, B.; Hristovski, K. Development and validation of RRLC-UV method for determination of chlorogenic acid in green coffee. *Acta Chromatogr.* **2020**, *32*, 34. [CrossRef]
20. Chen, X.; Mao, J.; Wen, F.; Xu, X. Determination of phenolic acids in botanical pharmaceutical products by capillary electrophoresis with chemiluminiscence detection. *Anal. Lett.* **2021**, *54*, 8861765. [CrossRef]
21. Xia, Z.; Sun, Y.; Cai, C.; He, Y.; Nie, P. Determination of chlorogenic acid, luteoloside and 3,5-O-dicaffeoylquinic acid in *Chrysanthemum* using near—Infrared spectroscopy. *Sensors* **2019**, *19*, 1981. [CrossRef]
22. Bounegru, A.V.; Apetrei, C. Simultaneous determination of caffeic acid and ferrulic acid using carbon nanofiber-based screen-printed sensor. *Sensors* **2022**, *22*, 4689. [CrossRef] [PubMed]

**Disclaimer/Publisher's Note:** The statements, opinions and data contained in all publications are solely those of the individual author(s) and contributor(s) and not of MDPI and/or the editor(s). MDPI and/or the editor(s) disclaim responsibility for any injury to people or property resulting from any ideas, methods, instructions or products referred to in the content.

Article

# pH and NaCl Optimisation to Improve the Stability of Gold and Silver Nanoparticles' Anti-Zearalenone Antibody Conjugates for Immunochromatographic Assay

Thasmin Shahjahan [1,†], Bilal Javed [1,2,3,*,†], Vinayak Sharma [2] and Furong Tian [1,2]

1. School of Food Science and Environmental Health, College of Sciences and Health, Technological University Dublin, D07 H6K8 Dublin, Ireland; furong.tian@tudublin.ie (F.T.)
2. Nano Lab, FOCAS Research Institute, Technological University Dublin, D08 CKP1 Dublin, Ireland
3. RELX Elsevier, D18 X6N2 Dublin, Ireland
* Correspondence: bilal.javed@tudublin.ie
† These authors contributed equally to this work.

**Abstract:** The aim of this research is to define optimal conditions to improve the stability of gold and silver nanoparticles' anti-zearalenone antibody conjugates for their utilisation in lateral flow immunochromatographic assay (LFIA). The Turkevich–Frens method was used to synthesise gold nanoparticles (AuNPs), which were between 10 and 110 nm in diameter. Silver nanoparticles (AgNPs) with a size distribution of 2.5 to 100 nm were synthesised using sodium borohydride as a reducing agent. The onset of AuNP and AgNP aggregation occurred at 150 mM and 80 mM NaCl concentrations, respectively. Stable Au and Ag nanoparticle–antibody conjugates were achieved at 1.2 mM of $K_2CO_3$ concentration, which corresponds to the pH value of ≈7. Lastly, the highest degree of conjugation between Au and Ag nanoparticles and anti-zearalenone antibodies was at 4 and 6 µg/mL of antibody concentrations. The optimisation of the conjugation conditions can contribute to better stability of nanoparticles and their antibody conjugate and can improve the reproducibility of results of bioreporter molecules in biosensing lateral flow devices.

**Keywords:** antibody conjugates; biosensing; bioreporter; lateral flow devices; nanoparticles aggregation; pH

**Citation:** Shahjahan, T.; Javed, B.; Sharma, V.; Tian, F. pH and NaCl Optimisation to Improve the Stability of Gold and Silver Nanoparticles' Anti-Zearalenone Antibody Conjugates for Immunochromatographic Assay. *Methods Protoc.* **2023**, *6*, 93. https://doi.org/10.3390/mps6050093

Academic Editors: Victoria Samanidou, Verónica Pino and Natasa Kalogiouri

Received: 18 August 2023
Revised: 21 September 2023
Accepted: 27 September 2023
Published: 3 October 2023

**Copyright:** © 2023 by the authors. Licensee MDPI, Basel, Switzerland. This article is an open access article distributed under the terms and conditions of the Creative Commons Attribution (CC BY) license (https://creativecommons.org/licenses/by/4.0/).

## 1. Introduction

The detection of zearalenone via lateral flow immunochromatographic assays plays a significant role in food safety and agriculture. Zearalenone is a mycotoxin, which is produced by certain fungal species and poses significant risks to both human and animal health when present in food and feed beyond a certain level. Rapid detection of zearalenone through LFIA provides a cost-effective, user-friendly, and on-site screening device for farmers, food processors, and regulatory authorities. It enables the timely identification of contaminated crops and products, which facilitates immediate interventions to mitigate health risks and economic losses. Moreover, the portability and simplicity of the assay make it an essential tool in monitoring and ensuring the safety of our food supply chain.

In the LFIA, the signal labels are generally conjugated with a detection agent, such as antibodies for the specific binding to the target analyte, like mycotoxins. This allows a visible signal to be generated by the recognition elements to indicate the presence of the antigen. The challenge in immobilising antibodies onto AuNPs or AgNOs is avoiding aggregation and ensuring that the antibodies are orientated correctly to maintain the functionality and the accessibility of their paratopes to conjugate with the antigens or analyte [1]. In order for the signal labels to precisely and accurately detect zearalenone, they must fulfil a range of criteria, which include having high stability, exhibiting little or no non-specific binding, being cost-effective, and forming reproducible and efficient conjugates without compromising the functionality and activity of the detection molecule [2].

To effectively use antibody–NP (Ab-NP) bioconjugates for biosensing, it is vital to develop robust and reliable techniques to ensure that the produced biosensor is reproducible, selective, and sensitive. An efficient bioconjugation approach must preserve the colloidal stability of the nanoparticles (NPs) while maintaining the capacity of Ab-NP bioconjugates to identify their target antigen [3]. Nanoparticles can be conjugated by physical adsorption, which is typically the preferred method for LFIA applications. It involves immobilising detection molecules onto noble metal nanoparticle's surfaces via hydrophobic, electrostatic interactions, hydrogen bonds, and Van der Waals forces [4]. The optimisation of this process can be achieved by testing different pH values near the isoelectric point of the binding molecule [5]. However, physical adsorption can lead to unreliable results due to the erroneous orientation of detection molecules, leading to the blocking of binding sites and poor reproducibility [6]. Controlling the pH between 7.5 and 8.5 can help control the orientation of antibodies and improve direct binding to citrate-stabilized NPs [7].

To determine the optimal quantity of antibody for adsorption, the flocculation test is conducted. This test identifies the minimum amount of antibodies needed to maintain the stability of NPs against salt-induced aggregation. Usually, 10% of NaCl is used for the flocculation test, which is followed by measuring the colour change in the colloidal solution or the absorbance of the solution at its λ max [8]. An optimal conjugate can be achieved by adding varying concentrations of antibodies to the AuNPs/AgNPs. If the conjugate aggregates upon contact with sodium chloride, it indicates that the mixture does not contain a sufficient number of conjugated antibodies, resulting in a colour change in the solution from red to blue [9,10].

Changes in the environment of AuNPs/AgNPs often result in the formation of aggregates. The term aggregate is used to refer to individual nanoparticles that have interacted with each other to form a larger super-structure without altering the shape or size of individual nanoparticles [11]. As maintaining the stability of conjugates in LFIA strips is crucial, it is essential to gain a deeper understanding of the nanoparticle conjugates for the effective optimisation of their performance in lateral flow assay. This can be achieved by characterising standard nanoparticles and conjugates using analytical techniques and measuring different parameters such as size, shape, zeta potential, absorbance, and optical density to monitor their stability [12].

The size of nanoparticles plays a crucial role in the sensitivity of LFIA. If aggregation occurs, the colour of gold nanospheres in suspension changes from wine-red to darker shades, affecting the intensity of the lines on the strip [13]. According to the study conducted by Sahoo and Singh [14], the size of nanoparticles can be controlled by adjusting parameters such as the concentration of sodium citrate, pH, and temperature. Nanospheres with diameters in the range of 20–40 nm are commonly used in optimising parameters for LFIA sensitivity, as larger nanoparticles can provide enhanced colour observation; however, they are less stable [14]. As aggregation is a challenge during conjugation, monitoring the size of standard nanoparticles as well as conjugates can aid in determining their aggregation state [15].

The aim of this project was to develop stable gold and silver nanoparticle anti-zearalenone antibody conjugates. It involved the synthesis and characterisation of AuNPs and AgNPs and their anti-zearalenone antibody conjugates. Three main studies were performed for optimising nanoparticle–antibody conjugate conditions. Various physical and optical parameters were measured using DLS, SEM, and UV-visible spectroscopy. These include absorbance, optical density, particle shape, particle size, polydispersity index (PDI), and zeta potential. In the first study, the nanoparticles were studied under various concentrations of NaCl in a 96-well plate to examine the aggregation behaviour of nanoparticles in an alkaline environment. A second study was performed to determine the optimum pH at which successful nanoparticle–antibody conjugation can be achieved. It involved the conjugation of nanoparticles with 2 µg/mL of antibodies under varying concentrations of $K_2CO_3$ to adjust the pH of the reaction mixture. The NaCl concentration at which the onset of nanoparticle aggregation occurred was added to determine the success of the conjugation process. The stability of the conjugates and the aggregation state were evaluated via the characterisation of conjugates. A third study was carried out where the concentration of $K_2CO_3$ and NaCl were kept constant, and the antibody concentration was varied to determine the concentration at which maximum nanoparticle–antibody conjugation can be achieved under a controlled environment. Therefore, this study provides a novel comparative systematic analysis to develop stable gold and silver nanoparticle–antibody conjugates to improve overall performance and to enhance the accuracy of LFIA test results.

## 2. Material and Methods

### 2.1. Materials

Gold (III) chloride trihydrate ($HAuCl_4 \cdot 3H_2O$), trisodium citrate, $AgNO_3$, $NaBH_4$, NaCl, $K_2CO_3$, and sodium hydroxide (NaOH) were purchased from Sigma-Aldrich, Dublin, Ireland. Anti-zearalenone [11C9] mouse monoclonal antibodies were procured from Abcam. Deionised water (DI) was produced by using an Elix Reference Water Purification System from Millipore, Ireland, and was used for the preparation of the solution. All the chemicals were used as received from the supplier without further purification or modification.

### 2.2. Gold Nanoparticles Synthesis

According to the procedure outlined by Turkevich et al. (1951), colloidal AuNPs were produced using the Turkevich–Frens technique [16]. A 250 mL glass beaker was filled with 250 µL of 50.5 mM $HAuCl_4$ stock solution and 94.75 mL of deionised water. The mixture was heated to a temperature of 100 °C and then placed on a hot stirrer. At a moderate rate of mechanical stirring, 5 mL of 1% trisodium citrate was added to the boiling solution at a rate of 1 mL/s. The solution started out faintly pink after two minutes and steadily became darker over the course of approximately eight minutes. After a deep wine-red colour appeared, the reaction was stopped by removing it from the water bath and allowed to cool down at room temperature.

### 2.3. Silver Nanoparticles Synthesis

AgNPs were generated using the procedure outlined in [17]. An amount of 30 mL of deionised water was used to dissolve 0.0023 g of $NaBH_4$ to create a final concentration of 0.002 M. $AgNO_3$ was dissolved in 10 mL of deionised water to a final concentration of 0.001 M using 0.0017 g of the reagent. The 0.002 M $NaBH_4$ solution was incubated for 20 min in an ice batch. The $NaBH_4$ solution was subsequently mixed with 0.001 M of $AgNO_3$ at a rate of 1 mL/s. A cold magnetic stirrer was used to vigorously stir the reaction mixture. The solution turned pale yellow following the addition of 2 mL of $AgNO_3$, and it eventually turned bright yellow after incorporating the entire quantity of $AgNO_3$.

### 2.4. Gold and Silver Nanoparticles Aggregation Study under the Influence of Salt

A total of 150 µL of colloidal gold/silver nanoparticles were loaded in a 96-well plate. An amount of 150 µL of deionized water was added to the first three wells as the control. Then, duplicate additions of 150 µL of NaCl at various concentrations (20, 40, 60, 80, 100, 180, 200, and 400 mM) were made. For approximately two minutes, the plate was left to incubate at room temperature. The colour changed after two minutes. The SpectroMax plate reader was used to quantify the OD and absorbance [12].

### 2.5. Investigating pH Changes by Using Various $K_2CO_3$ Concentrations against Silver and Gold Nanoparticles

A micropipette was used to add 150 µL of colloidal gold/silver nanoparticles into the wells of a 96-well plate. An amount of 150 µL of deionized water was placed into the first three wells of the first row for control 1. Then, for control 2, 150 µL of deionised water and 2 µg/mL of antibody solution were pipetted into the first three wells of the second row. To achieve the following concentrations—0.0 mM, 0.2 mM, 0.4 mM, 0.6 mM, 0.8 mM, 1.0 mM, and 1.2 mM—and a final volume of 100 µL, various amounts of a 10 mM $K_2CO_3$

stock solution were then added to subsequent rows. To each well, 2 μg/mL of the antibody solution was added. To ensure the antibodies were evenly distributed throughout the nanoparticle solution, the mixtures were reverse pipetted. For approximately 20 to 25 min, the plate was set aside at room temperature. Next, to the Au and AgNPs, 200 mM and 80 mM NaCl, respectively, were added. The OD and absorbance were measured using the SpectroMax plate reader once colour change occurred [12].

*2.6. Study of Silver and Gold Nanoparticles at Different Antibody Concentrations*

A micropipette was used to add 150 μL of colloidal gold/silver nanoparticles into the wells of a 96-well plate. An amount of 150 μL of deionized water was added into the first three wells of the first row for control 1. Then, for control 2, 150 μL of deionized water and 2 ug/mL of antibody solution were pipetted into the first three wells of the second row. With the exception of the control wells, 100 μL of $K_2CO_3$ optimal concentration was introduced to each well. The antibody solution was then added in different concentrations. Amounts of 2 and 4 μg/mL were employed for AuNPs. For AgNPs, antibodies at concentrations of 2, 4, and 6 μg/mL were used. To ensure that the antibodies were evenly distributed throughout the nanoparticle solution, the mixtures were reverse pipetted. For approximately 20 to 25 min, the plate was left to rest at room temperature. Then, for the addition of Au and Ag NPs, 200 mM and 80 mM NaCl were added, respectively. The OD and absorbance were measured using the SpectroMax plate reader once colour change occurred.

*2.7. Particle Characterisation*

On a Malvern Instrument Nano-Zetasizer, DLS measurements were performed. In order to prepare the samples, 900 μL of deionised (DI) water and 100 μL of the nanoparticle colloidal sample from Section 2.2 or Section 2.3 were mixed in a disposable plastic cuvette. Prior to the measurement, the samples were allowed to adjust to 25 °C for approximately 2 min. After three successive runs with each sample, graphs of the size distribution by intensity were analysed.

A pipette was utilised to transfer 800 μL of the sample from the disposable cuvette to a Malvern Panalytical Folded Capillary Zeta Cell for ZP measurement. The UV-visible spectrophotometer did not require dilution to measure absorbance. One millilitre of the sample was added to a small cuvette before being placed in the sample container. Transmission Electron Microscopy (TEM) was used to investigate the morphology and size of AuNPs and AgNPs [9,18–20].

**3. Result and Discussion**

*3.1. Characterisation of Gold and Silver Nanoparticles*

The synthesis of AuNPs and AgNPs was achieved using trisodium citrate as the reducing agent for AuNPs and silver nitrate for AgNPs. The reduction reactions were conducted under carefully optimised conditions including controlled temperature and reaction time to facilitate the nucleation and growth of stable nanoparticles. The particle spectra and images are presented in Figure 1.

Figure 1a,b show the UV-visible spectra of the synthesised silver and gold NPs exhibit distinct absorption peaks in the 300 nm to 800 nm spectral range. Evident within these spectra are well-defined characteristic plasmon resonance peaks observed at 520 nm for AuNPs and 400 nm for AgNPs. These plasmon resonance peaks indicate the collective oscillation of electrons on the nanoparticle surface. This confirms the successful formation of colloidal Ag and Au nanoparticles.

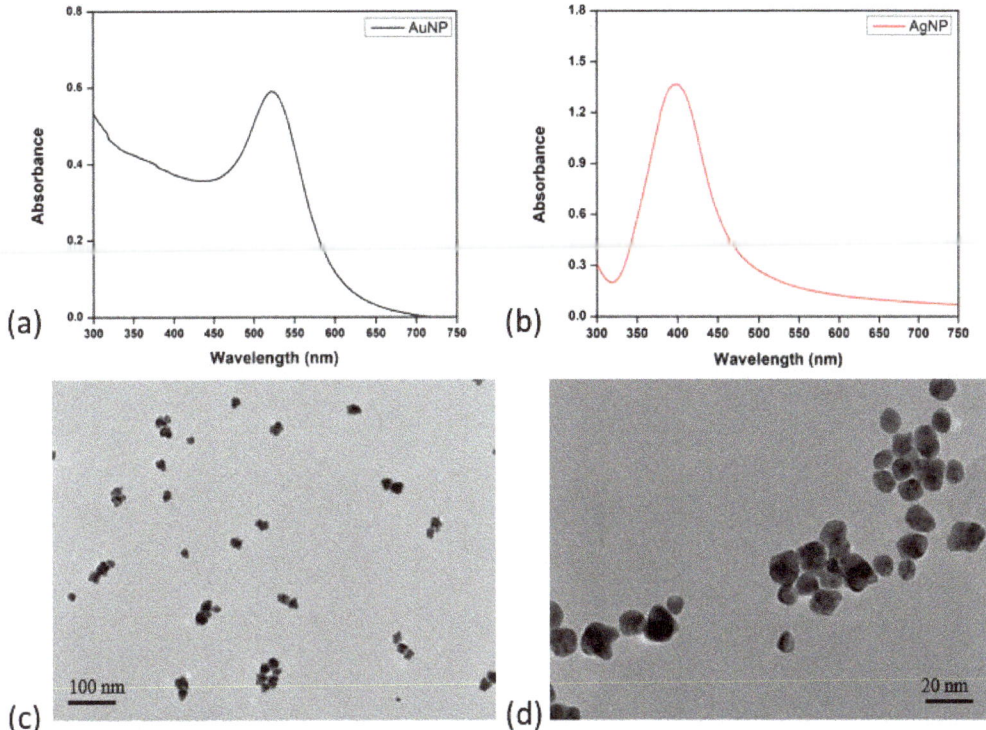

**Figure 1.** Characterisation of gold and silver nanoparticles. (**a**) The UV-visible spectra of AuNPs, (**b**) the UV-visible spectra of AgNPs, (**c**) a TEM image of AuNPs, and (**d**) a TEM image of AgNPs.

The TEM images revealed that both nanoparticles display a spherical morphology with a relatively narrow size distribution (Figure 1c,d). AgNPs exhibited diameters ranging from 17.6 nm to 25.8 nm, while AuNPs demonstrated an average diameter spanning from 5.95 nm to 11.9 nm. When conjugating the AuNPs and AgNPs with antibodies, the spherical morphology of the synthesised AuNPs and AgNPs may be advantageous. This is primarily attributed to the potential enhancement in achieving a well-orientated distribution of antibodies across the nanoparticle surface. Consequently, this may potentially enhance both the binding efficiency and sensitivity [21].

The size of nanoparticles holds the potential to significantly influence the overall performance of the application in lateral flow assay [22]. Dynamic Light Scattering (DLS) was employed to determine the hydrodynamic size distributions of the nanoparticles. The obtained size distribution by intensity profiles, which indicates that the AuNPs exhibited a size range of 9 nm to 120 nm, while the AgNPs were observed to range from 2.5 nm to 120 nm. These findings denote the presence of a well-dispersed population. The range suggests that the nanoparticles are predominantly polydisperse. AgNPs have a relatively broader range of hydrodynamic sizes compared to AuNPs. The introduction of larger nanoparticles could be advantageous. According to their findings, increasing the nanoparticle size to as high as 115 nm resulted in a significant drop in the limit of detection (LOD) [23]. The polydispersity index (PDI) and Zeta Potential (ZP) values are shown in Table 1.

Au and Ag nanoparticles had zeta potentials of $-26.3 \pm 4.6$ mV and $-20.07 \pm 0.5$ mV, respectively, showing strong negative surface charges. They are relatively stable as zeta potential values other than $-30$ mV to $+30$ mV are deemed stable [24]. Furthermore, the zeta potential of the AuNPs was comparatively stronger than that of the AgNPs, indicating better stability.

**Table 1.** ZP and PDI of standard gold and silver nanoparticles measured using DLS.

|  | Gold Nanoparticles | Silver Nanoparticles |
| --- | --- | --- |
| ZP (mV) | −26.3 ± 4.6 | −20.07 ± 0.5 |
| PDI | 0.209 | 0.564 |

It was reported that PDI greater than 0.7 is an indication of aggregated nanoparticles in the solution [25]. The current result shows a PDI value of 0.2 for the AuNPs, indicating a narrow size distribution, meaning that the majority of the particles are similar in size. On the other hand, a PDI value of 0.5 was obtained for the AgNPs, which is typically acceptable and suggests a broader size distribution.

*3.2. Analysis of Gold and Silver Nanoparticles Aggregation Behaviour in Differing Alkaline Environments*

Prior to antibody conjugation, the stability of the synthesised Au and Ag nanoparticles was investigated in NaCl concentrations ranging from 20 mM to 400 mM. The aim was to investigate the effects of different NaCl levels on nanoparticle stability. For accuracy, the experiment used a 96-well plate format with triplicate samples (Figure 2).

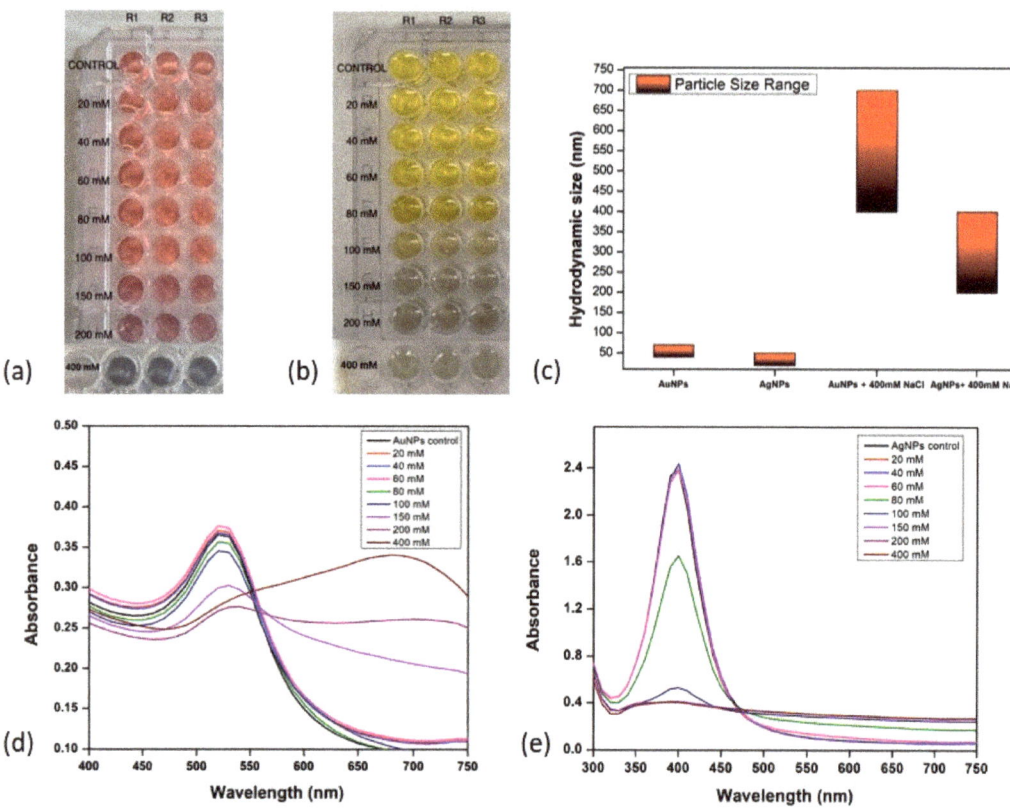

**Figure 2.** Gold and silver nanoparticles aggregation behaviour in differing alkaline environments. (**a**) Image of AuNP with series of NaCl solution in the 96 well plate. (**b**) Image of AgNP with series of NaCl solution in the 96 well plate. (**c**) Histogram of the hydrodynamic size of AuNPs and AgNPs with and without 400 mM NaCl. (**d**) The spectra of AuNPs under different concentrations of NaCl. (**e**) The spectra of AgNPs under different concentrations of NaCl.

Figure 2 depicts the aggregation behaviour of AuNPs (Figure 2a,d) across various NaCl concentrations. At NaCl concentrations of 20 mM, 40 mM, 60 mM, 80 mM, and 100 mM, AuNPs displayed no aggregation while retaining their typical brilliant red colour and stability. However, around 150 mM, a shift from red to pink-purple occurred, suggesting that the physical and chemical characteristics had changed in response to unfavourable alkaline conditions. At a NaCl concentration of 150 mM, the stability of AuNPs was disturbed, resulting in aggregation. The observed grey colour indicates that saturation was achieved at 400 mM. Similar trends were observed for AgNPs, with slight variations. The onset aggregation of AgNPs was observed at 80 mM of NaCl, characterised by a transition from brilliant yellow to light grey colouration (Figure 2b,e). Following that, saturation at 150 mM of NaCl was observed, resulting in the development of dark grey nanoparticles. These varied observations together highlight the complex relationship between nanoparticle stability and NaCl concentration.

For additional insight into the behaviour of the AuNPs in the alkaline environment, the optical density (OD) was measured at 520 nm and 630 nm to monitor changes in their SPR properties as a function of NaCl concentration. The OD measurements obtained at 520 nm and 630 nm were plotted on a graph, which is shown in Figure 3.

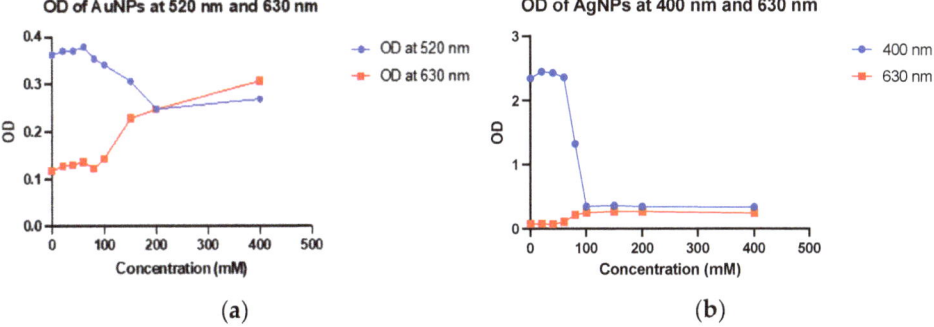

**Figure 3.** Optical density (OD) of AuNP and AgNP at varying NaCl Concentrations. (**a**) OD of AuNP was taken at 520 nm and 630 nm. OD decreases at 520 nm and increases at 630 nm; (**b**) OD of AgNP at varying NaCl Concentrations. OD decreases at 400 nm and increases slightly at 630 nm. Note: 0 mM (control), 20. NaCl Concentrations of 0 mM (control), 20 mM, 40 mM, 60 mM, 80 mM, 100 mM, 150 mM, and 400 mM were used.

The OD at 520 nm decreased slightly at NaCl concentrations 20 mM to 100 mM, while at NaCl concentrations 150 mM to 400 mM, a more significant decrease was observed (Figure 3a). This indicated a decrease in the concentration of dispersed nanoparticles as the NaCl concentration increased due to increased aggregation. The SPR of the nanoparticles slightly shifted to a longer wavelength at NaCl concentrations of 20 to 100 mM. These shifts were more significant as the NaCl concentration increased beyond 100 mM; hence, a significant decrease in OD was observed. This is consistent with the red shift in the SPR peak observed in Figure 2d. These shifts in SPR occurred as a result of the interaction of Na+ and Cl- ions with the surface charge of nanoparticles, with stronger interactions occurring at higher NaCl concentrations. Conversely, at 630 nm, the OD increased with increasing NaCl concentration (Figure 3b). This was clearly due to the shift in the SPR of nanoparticles to longer wavelengths. The increase in OD became more significant at 150 mM, which is the concentration at which the onset of nanoparticle aggregation occurs, as determined by the corresponding SPR peak on the UV-visible absorption spectra.

The hydrodynamic sizes of both gold and silver nanoparticles (without NaCl influence) and with NaCl were measured using dynamic light scattering analysis (Figure 2c). The data reveals the size of AuNPs in the control sample was in the range of 40–70 nm and of AgNPs was 20–40 nm. After the addition of NaCl at the concentration of 400 mM, a change in colour was observed, which shows the aggregation of both AuNPs and AgNPs. The aggregation was confirmed by checking the hydrodynamic size, zeta potential, and PDI values of dispersions of AuNPs and AgNPs. The size was found to be in the range of 450–700 nm for AuNPs and 200–400 nm in the case of AgNPs. PDI values were also found to be close to 1, along with zeta potential close to zero, indicating unstable and aggregated dispersions. The colour changed to grey, indicating that high NaCl concentration led to aggregation. Hence, for further experiments, the concentration of NaCl used was 150 mM for AuNPs and 80 mM for AgNPs.

*3.3. Influence of Colloidal Solution pH on Antibody Conjugation to Gold and Silver Nanoparticles*

The effect of varying pH levels on the interaction between the nanoparticles and antibodies was investigated. $K_2CO_3$ was employed to adjust the pH, with concentrations ranging from 0.2 mM to 1.2 mM. To verify the conjugation, a suboptimal NaCl concentration of 150 mM was introduced for AuNPs and 80 mM for AgNPs. The study aimed to identify the optimal pH for antibody–nanoparticle (Ab-NP) conjugation. The pH of colloidal Au and Ag nanoparticle solutions was regulated using $K_2CO_3$, while antibody concentration remained constant at 2 µg/mL. The details of the $K_2CO_3$ concentrations used to adjust the pH of the colloidal AuNPs and AgNPs solutions are provided in Figure 4a,c, respectively. Figure 4 shows the effect of varying pH levels maintained via $K_2CO_3$ on the interaction between the nanoparticles and antibodies. Figure 4a,c illustrates visually monitored solutions for colour changes in the 96-well plate.

The image in Figure 4a shows the plate used for the colourimetric assay of Ab-AuNP conjugates. Colour changes occurred at concentrations of 0.2, 0.4, 0.6, 0.8, and 1 mM, signifying nanoparticle aggregation after NaCl addition. This indicated unsuccessful antibody–AuNP conjugation due to suboptimal pH. Lower $K_2CO_3$ concentrations created a low-pH environment, causing electrostatic bridging between antibodies and negatively charged citrate-capped nanoparticles, resulting in aggregation. At 1.2 mM $K_2CO_3$, corresponding to the pH 7.65, the nanoparticles exhibited minimal colour change compared to the control, indicating successful antibody coating onto nanoparticle surfaces.

To investigate the effect of pH on AuNP optical properties, nanoparticle absorbance was measured using UV-visible spectroscopy (300–750 nm). Figure 4b depicts the obtained absorbance spectra. SPR peaks at around 520 nm, similar to standard AuNPs, were identified. Control 1 (AuNPs and deionised water) and Control 2 (AuNPs, deionised water, and antibodies) exhibited no changes in SPR, optimal antibody orientation on the nanoparticle surface, and stability. As the reaction pH decreased, slight redshifts appeared, suggesting augmented nano-moiety size. Changes in interparticle distances and dielectric constants of the surrounding medium are two causing factors. Lower pH resulted in larger, less defined SPR peaks, indicating destabilisation and aggregation of nanoparticle conjugates. The SPR peak closely mirrored the control peaks at 1.2 mM $K_2CO_3$, indicating stable antibody–AuNP conjugation in solution.

Figure 4c is an image of the 96-well plate employed for examining AgNPs. Control 1 contains AgNPs and deionized water, and Control 2 also includes antibodies. AgNPs followed a similar aggregation pattern to AuNPs, aggregating at 0.2, 0.4, 0.6, 0.8, and 1 mM, evident by colour shifts from bright yellow to pale yellow to light grey. Retention of bright yellow colour at 1.2 mM, which corresponds to pH 7.95, suggested optimal $K_2CO_3$ concentration for AgNPs-Ab conjugation.

The particle size distribution for control 1 (AuNPs and deionised water) was reported between 10 to 140 nm, while the values changed from 8 to 90 nm for control 2 (AuNPs, deionised water, and antibodies). At 1.2 mM $K_2CO_3$, the particle size distribution increased from 10 to 410 nm following conjugation, which shows an increase in the nanoparticle standard size and indicates the effective coverage of the nanoparticle's surface with the antibodies. Figure 4d presents UV-visible absorbance spectra of Ab-AgNP complexes at varied $K_2CO_3$ concentrations. The SPR peaks are positioned around 400 nm. However, a significant change in peak shape and intensity is observed at low concentrations of

$K_2CO_3$, indicating aggregation stability. Conjugates showed a significant absorbance drop at 400 nm for $K_2CO_3$ concentrations 0.2 mM to 1.2 mM, with 1.2 mM exhibiting a slightly smaller yet well-defined peak, suggesting stability. Table 2 illustrates the zeta potential (ZP) and PDI of AuNPs in control 1, control 2, and at 1.2 mM of $K_2CO_3$ concentration.

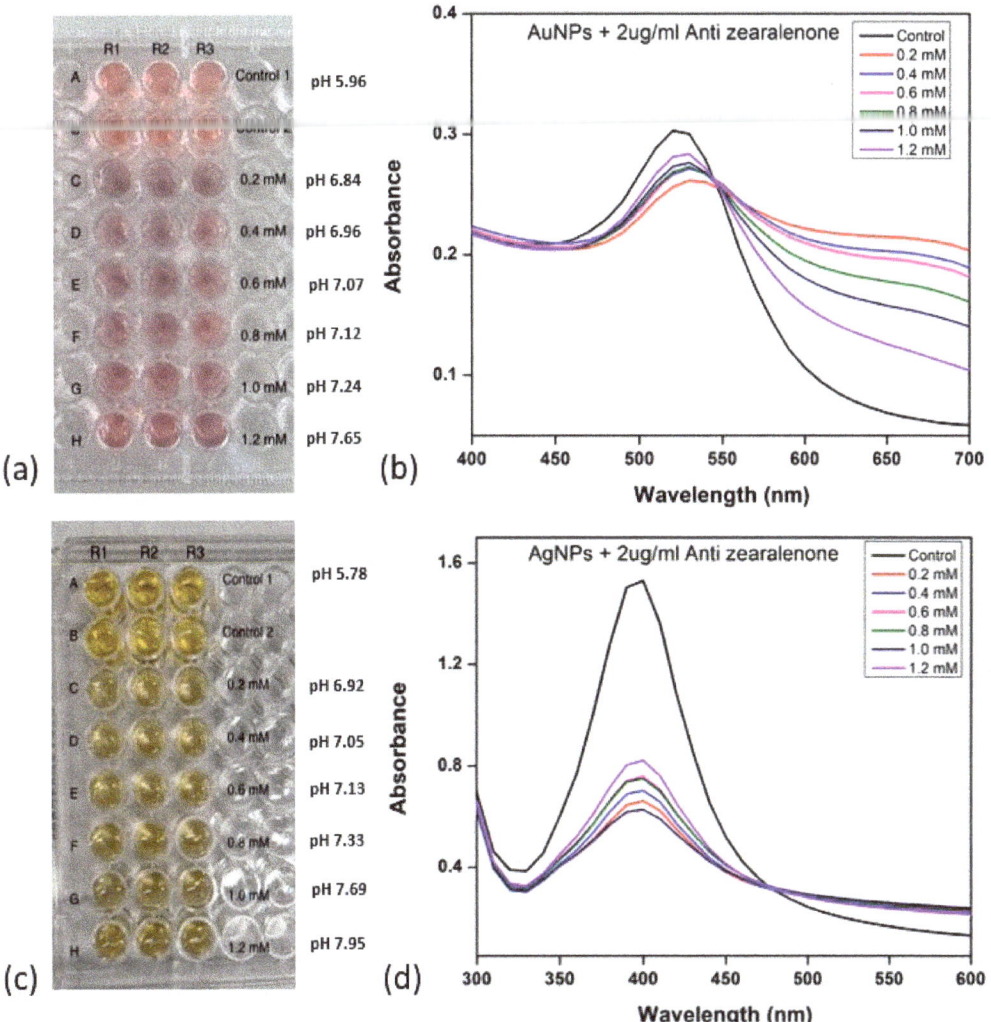

**Figure 4.** Gold and silver nanoparticle aggregation behaviour in differing alkaline environments. (**a**) Image of AuNP with the series anti-zearalenone solution in the 96-well plate. (**b**) The spectra of AuNP at varied anti-zearalenone concentrations. (**c**) Image of AgNP with the series anti-zearalenone solution in the 96-well plate. (**d**) The spectra of AgNP at varied anti-zearalenone concentrations.

**Table 2.** ZP and PDI of gold nanoparticles in controls and at 1.2 mM of $K_2CO_3$ concentration.

| $K_2CO_3$ Concentration | ZP (mV) | PDI |
|---|---|---|
| Control 1 | −33.4 ± 8.2 | 0.229 |
| Control 2 | −30.9 ± 10.6 | 0.329 |
| 1.2 mM | −23.2 ± 1.5 | 0.347 |

Control 1, control 2, and Ab-AuNP at 1.2 mM $K_2CO_3$ all had significantly negative zeta potential values of $-33.4 \pm 8.2$ mV, $-30.9 \pm 10.6$ mV, and $-23.2 \pm 1.5$ mV, indicating highly stable conjugates in Table 2. A slight reduction in ZP at 1.2 mM of $K_2CO_3$ suggested a modest decrease in surface charge, presumably due to pH-driven changes in ionic strength influencing electrostatic interactions between nanoparticles in the surrounding medium. The Polydispersity Index (PDI) values were 0.229, 0.329, and 0.347 (Table 2).

Notably, the conjugates at optimum pH had a larger PDI, indicating a greater variance in nanoparticle size and enhanced solution heterogeneity [26,27]. This corresponds to the increased hydrodynamic diameter range of single AuNP–antibody conjugates. Similar results were obtained for the AgNPs and their antibody conjugates at the 1.2 mM concentration of $K_2CO_3$ (Table 3).

**Table 3.** ZP and PDI of silver nanoparticles in controls and at 1.2 mM of $K_2CO_3$ concentration.

| $K_2CO_3$ Concentration | ZP (mV) | PDI |
| --- | --- | --- |
| Control 1 | $-32.7 \pm 0.26$ | 0.179 |
| Control 2 | $-31.5 \pm 5.20$ | 0.166 |
| 1.2 mM | $-33.6 \pm 1.11$ | 0.312 |

To gain further insights into antibody and nanoparticle behaviour across varying pH, OD measurements were taken at 530 nm and 630 nm on different concentrations of $K_2CO_3$ (Figure 5).

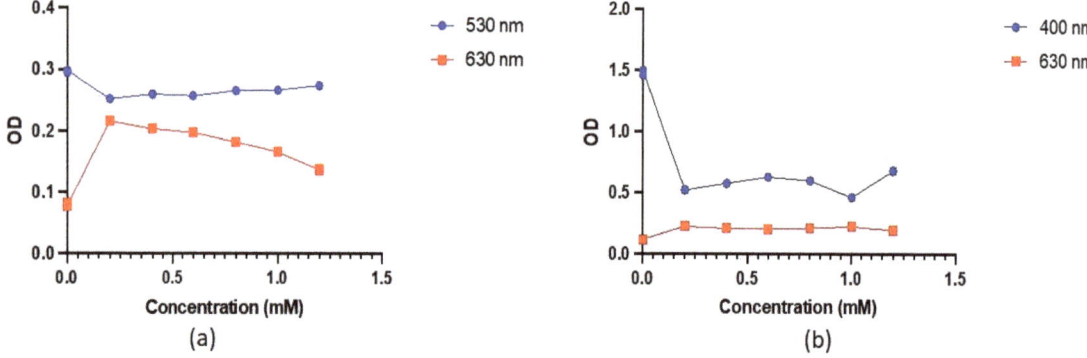

**Figure 5.** Optical density (OD) of AuNPs and AgNPs at varying $K_2CO_3$ concentrations. (**a**) OD of AuNP was taken at 530 nm and 630 nm. OD decreases at 520 nm and increases at 630 nm. (**b**) Optical density (OD) of AgNP at $K_2CO_3$ concentrations. Note: 0 mM (control at $K_2CO_3$ concentrations 0.2 mM, 0.4 mM, 0.6 mM, 0.8 mM, 1.0 mM, and 1.2 mM.

At 530 nm, a decrease in OD values was observed, with a notable decrease from control 1 to control 2. As pH increased, OD increased consistently. This pattern is mirrored in the OD measurements at 630 nm—a significant increase from control 1 to control 2—followed by a steady decrease with rising pH. The variations in pH influence the surface charge of the nanoparticles, impacting the adsorption of antibodies. These changes in surface properties can impact the SPR phenomenon, causing shifts towards longer wavelengths as a result of more prevalent repulsive interactions between particles. This is consistent with the redshift in SPR peaks at low $K_2CO_3$ concentrations observed in the UV-visible spectra. The behaviour observed at both wavelengths suggests that antibody–nanoparticle interactions are closely connected to pH-induced changes in surface charge and nanoparticle stability (Figure 5a).

OD measurements of AgNPs-Ab conjugates in Figure 5b reveal a decrease at lower pH and an increase at 1.2 mM at 400 nm. Contrastingly, variations at 630 nm are minimal.

For conjugates at 1.2 mM $K_2CO_3$ concentration, the DLS data demonstrate a broad size distribution of 10 to 400 nm, highly negative ZP of $-33.6 \pm 1.11$ mV, and slightly higher PDI of 0.312 compared to the controls (Table 3). These findings are similar to that of AuNPs antibody conjugates in this study, providing evidence of successful antibody binding for stable conjugates at 1.2 mM of $K_2CO_3$ concentration.

Kasoju et al. [12] found that the onset of nanoparticle aggregation occurred at 40 mM and reached saturation at 80 mM. This differs from the findings of this study, as the nanoparticles began aggregating at 80 mM and reached saturation at 400 mM. This means that the nanoparticles in this study have a higher stability, are more resistant to aggregation, and require higher NaCl concentrations to induce aggregation compared to the AuNPs synthesised by Kasoju et al. [12]. Additionally, the higher concentration range for saturation suggests that AuNPs have a comparatively higher sensitivity to changes in NaCl concentration. In this study, the AuNPs size ranged between 9 and 120 nm, while the AuNPs used by Kasoju et al. [12] were approximately 18 to 20 nm. The difference in aggregation behaviour can be attributed to the size distribution of the particles, as particle size plays an important role in their stability and sensitivity to aggregation [28]. As the nanoparticles in this study were larger, they exhibited higher stability and began aggregating at a relatively higher NaCl concentration. This is advantageous for the use of these conjugates as they would be less likely to aggregate during migration on the sample pad, therefore ensuring better dispersion and enhanced signal generation for the lateral flow test strips. However, it is important to note that if the size of NPs is too large and excessively high NaCl concentrations are required to reach saturation, it could result in reduced assay sensitivity [29,30].

### 3.4. Stability Study of Gold and Silver Nanoparticles at Varying Antibody Concentrations under Controlled pH and NaCl Conditions

This study aimed to establish the optimal antibody concentration for enhancing nanoparticle surface binding and subsequently amplifying signal output. The antibody concentration was increased in 2 µg/mL increments, and pH was controlled using 1.2 mM $K_2CO_3$ concentration. The experimentation was conducted within a 96-well plate setup, with a control comprising nanoparticles, deionised water, and varying antibody concentrations. The arrangement of nanoparticles in the plate is illustrated in Figure 6.

AuNPs and AgNPs were conjugated with 2, 4 and 6 µg/mL of anti-zearalenone antibodies at the optimised pH value of $\gtrsim 7.5$. After adding the required concentration of antibodies and nanoparticles, NaCl solution was added. Nanoparticle conjugates exhibited minor colouration changes when exposed to suboptimal NaCl concentrations of 150 mM (AuNPs) and 80 mM (AgNPs). For further evaluation, OD measurements were taken at 520 nm, and variations in the OD of AuNP conjugates were evident (Figure 6b). Specifically, at 150 mM NaCl and 4 µg/mL of antibody concentration, the OD of nanoparticles (0.250) closely resembled the control sample's OD (0.298). This suggests that the SPR was similar under these conditions, indicating successful antibody attachment. At 630 nm, a similar pattern was observed, suggesting optimal conjugation at 4 µg/mL of antibody concentration at 150 mM of NaCl. Similarly, in the case of AgNPs, the OD measurements were taken at 400 nm. However, the extent of aggregation can be regarded as acceptable, as the Ab-AuNP and Ab-AgNP conjugates retained a substantial degree of the nanoparticles' original pinkish-red and bright yellow colour, respectively.

The influence of the varying antibody concentrations on the SPR of nanoparticle conjugates was evaluated via absorbance measurements employing UV-visible spectroscopy. In the UV-visible spectra of AuNPs-Ab conjugates (Figure 6b), the SPR peak of the control conjugates is distinct and well-defined. Conversely, the peaks corresponding to the AuNPs conjugated with 2 and 4 µg/mL of antibody concentrations are relatively broader and moved toward the higher wavelength, suggesting successful conjugation of AuNPs with the antibodies.

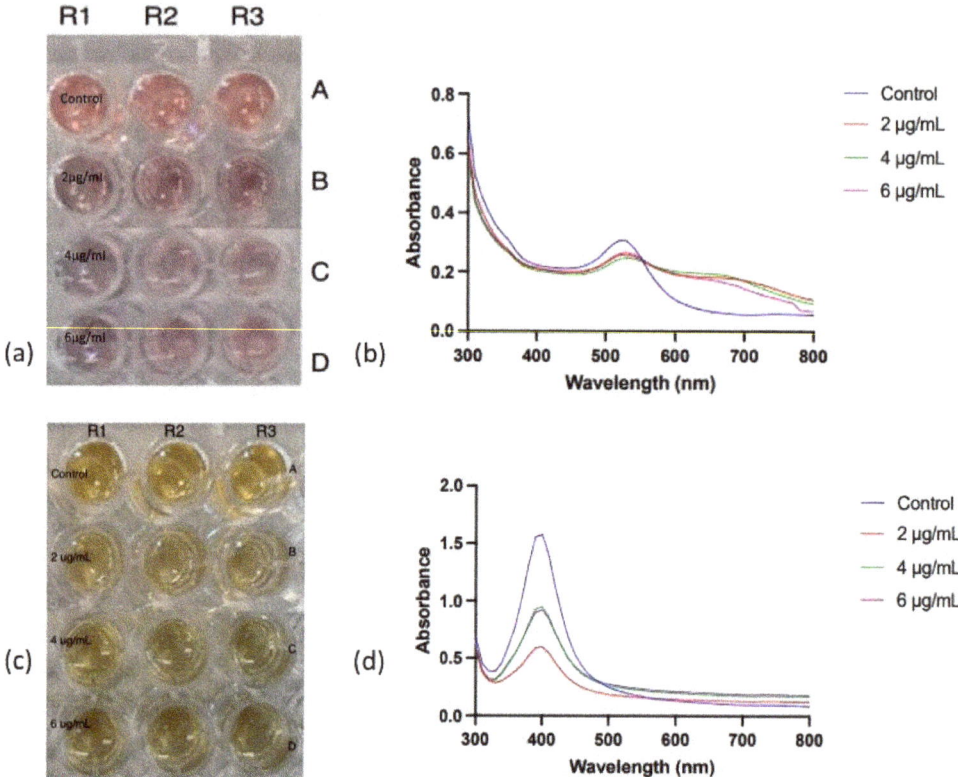

**Figure 6.** Colourimetric assay of AuNPs at varying antibody concentrations in an alkaline environment. (**a**) Row A is the control, which contains AuNPs, DI water, and antibodies. In rows B–D, the concentration of NaCl and $K_2CO_3$ was kept constant at 200/150 mM and 1.2 mM, respectively. The varying antibody concentrations were 2 and 4 µg/mL. R1, R2, and R3 are replicates. (**b**) A UV-visible spectrum of AuNPs and various concentrations of anti-zearalenone antibody conjugates. (**c**) Row A is the control, which contains AgNPs, DI water, and antibodies. In rows B–D, the concentration of NaCl and $K_2CO_3$ was kept constant at 200/150 mM and 1.2 mM, respectively. The varying antibody concentrations were 2, 4, and 6 µg/mL. R1, R2, and R3 are replicates. (**d**) A UV-visible spectrum of AgNPs and various concentrations of anti-zearalenone antibody conjugates.

In the case of AgNP-Ab conjugates (Figure 6c), the UV visible absorbance spectra display SPR peaks at 400 nm (Figure 6d). For conjugates at 2, 4, and 6 µg/mL of antibody concentrations, SPR peak intensities are lower than that of the control. Notably, 2 µg/mL exhibits the lowest intensity, while 4 µg/mL and 6 µg/mL share slightly higher intensities. An optimal antibody concentration for AgNP conjugation may not be 2 µg/mL due to the lowest SPR intensity, while 4 µg/mL and 6 µg/mL show comparable intensities, suggesting either could be optimal. Since the study's goal is to maximise the nanoparticle-to-antibody conjugation, 6 µg/mL emerges as the optimal antibody concentration.

The OD measurements for AgNP conjugates revealed no distinct trend at 400 nm and 630 nm, likely due to comparable surface plasmon resonances among conjugates at varying antibody concentrations under controlled pH and salt conditions. Notably, control containing only nanoparticles and antibodies exhibited significantly higher OD at 400 nm compared to conjugates with varying antibody concentrations. This divergence implied differing SPR bands, possibly due to non-optimised conditions for these control conjugates.

The hydrodynamic diameter of AuNP-Ab conjugates at 2 µg/mL in 0 mM NaCl, 2 µg/mL in 150 mM NaCl, and 4 µg/mL in 150 mM NaCl were 10–90 nm, 150–450 nm, and 150–400 nm, respectively. These measurements indicated that the conjugates exhibited larger sizes compared to the control. The same pattern was observed for AgNP-Ab conjugates where the size distribution of the control and conjugates at 2 µg/mL, 4 µg/mL, and 6 µg/mL ranged from 1 to 110 nm, 1–500 nm, 150–350 nm, and 195–500 nm, respectively. These observations strongly demonstrate the effective binding of antibodies to nanoparticles within the controlled pH environment. ZP and PDI of AuNPs at varying NaCl and Ab concentrations are detailed in Tables 4 and 5.

Table 4. ZP and PDI of gold nanoparticles at varying NaCl and Ab concentrations.

| NaCl and Ab Concentration | ZP (mV) | PDI |
|---|---|---|
| Control | −30.9 | 0.329 |
| 150 mM NaCl and 2 µg/mL Ab | −0.529 ± 0.54 | 0.362 |
| 150 mM NaCl and 4 µg/mL Ab | −32.5 ± 4.23 | 0.358 |
| 150 mM NaCl and 6 µg/mL Ab | −30.53 ± 1.80 | 0.411 |

Table 5. ZP and PDI of silver nanoparticles at varying NaCl and Ab concentrations.

| AgNPs (NaCl + Ab) | ZP (mV) | PDI |
|---|---|---|
| Control | −31.5 ± 0.72 | 0.166 |
| 150 mM NaCl and 2 µg/mL Ab | −22.9 ± 0.16 | 0.283 |
| 150 mM NaCl and 4 µg/mL Ab | −32.3 ± 0.63 | 0.345 |
| 150 mM NaCl and 6 µg/mL Ab | −36.01 ± 2.11 | 0.404 |

ZP measurements indicated comparable surface charges between control and antibody conjugates at 4 µg/mL of antibody concentration in 150 mM of NaCl, with values of −30.9 mV and −30.53 ± 1.80 mV, respectively. These highly negative readings denote substantial stability. In contrast, nanoparticle conjugates at 2 µg/mL of antibody in 150 mM of NaCl exhibited markedly lower negative ZP value of −0.529 ± 0.54 mV. This weak ZP signifies insufficient antibody coverage on nanoparticle surfaces, causing interactions with NaCl ions and the formation of large and destabilised nanoparticle aggregates. The low ZP aligns with DLS data, revealing larger particle sizes for the sample with 2 µg/mL of antibody concentration in 150 mM of NaCl. Additionally, the highest PDI appeared at 4 µg/mL of antibody in 150 mM of NaCl, indicative of increased heterogeneity and broad size distribution in these conjugates.

These findings effectively underscore that a more efficient conjugation process can be achieved at 4 µg/mL of antibody concentration in 150 mM NaCl with 1.2 mM of $K_2CO_3$ concentration. A similar study was conducted by Kasoju et al. [12], where AuNPs stability was examined using varying NaCl concentrations ranging from 20 mM to 200 mM. The aggregation pattern observed by the researcher was identical to the pattern observed in this study, where the nanoparticles at low NaCl concentration are stable and exhibit no colour change. Contrarily, at high concentrations of NaCl, the physical stability of the nanoparticles is affected, resulting in the formation of aggregates. The ZP of AgNP-Ab conjugates at 2 and 4 µg/mL of antibody concentrations were 0.329 ± 0.16 mV and 0.403 ± 0.639 mV, respectively. These lowly positive values typically suggest the potential destabilisation of conjugates in the reaction mixture. However, this contradicts the well-defined SPR peaks and comparable OD values, both indicative of conjugate stability. The discrepancy may stem from potential contamination during reaction mixture preparation or the conjugates' quality compromised due to prolonged room temperature exposure during analysis. The conjugates might have lost stability, possibly due to antibody sensitivity to temperature fluctuations in the environment and their requirement for low temperatures to preserve stability.

Conversely, in AgNPs at 6 µg/mL of concentration, the conjugates displayed a notably negative ZP value of −36 ± 2.11 mV, underscoring robust stability under optimised pH and salt conditions. This strongly implies enhanced temperature resistance compared to the 2 µg/mL and 4 µg/mL of antibody concentration conjugates, indicative of enhanced shelf life and sensitivity for biosensing. Hence, the optimal antibody concentration for AgNPs-Ab conjugation is 6 µg/mL. Bélteky et al. [31] performed a comparable study on AgNPs to determine the effect of NaCl on the colloidal stability of AgNP suspension. The results of the study also concluded that the SPR of the AgNPs declined with increasing NaCl concentration. The reasoning behind this is that the relative permittivity of the surrounding medium is increased with increasing NaCl concentration, thereby affecting the SPR [32]. Moreover, Bélteky et al. [31] observed changes in SPR at 10 mM and 50 mM of NaCl for AgNPs that were 10 nm and 20 to 50 nm, respectively, while in this study, this pattern was observed at 80 mM NaCl for AgNPs that are 2.5 to 120 nm. Therefore, this indicated that higher NaCl concentrations may be required to induce mild aggregation of larger nanoparticles.

Lou et al. [19] conducted a comparable investigation to determine an appropriate $K_2CO_3$ and antibody concentration for developing stable antibody-labelled AuNPs. The results demonstrated that the stability of the conjugates improved as the antibody concentration increased. This was because the nanoparticles were saturated by the antibodies, and hence the conjugates were less sensitive to NaCl and $K_2CO_3$ in the reaction mixture, preventing aggregation. A similar pattern was also observed in this study, where AuNP and AgNP conjugates were stable when the antibody concentration increased from 2 µg/mL to 4 µg/mL and 6 µg/mL, respectively. It was also found that a higher pH value is required for conjugating smaller-sized nanoparticles with antibodies compared to larger ones. This contrasted with the results of this study, as a higher pH value was required for developing stable AuNP and AgNP conjugates using Au and Ag NPs with a large hydrodynamic diameter of up to 120 nm.

## 4. Conclusions

In summary, gold and silver nanoparticles' anti-zearalenone antibody conjugates were successfully optimised. Standard gold and silver nanoparticles were synthesised using trisodium citrate and sodium borohydride, respectively, as reducing agents. AuNPs and AgNPs exhibited SPR peaks at 520 nm and 400 nm on the UV-visible absorbance spectra, respectively, which are the characteristic peaks of the nanoparticles. SEM analysis revealed that the AuNPs and AgNPs have a spherical morphology and are estimated to be between 5.78 nm to 11.8 nm and 30 nm to 150 nm, respectively. DLS analysis further confirmed the size distribution of AuNPs and AgNPs to be 9–120 nm and 2.5–120 nm, respectively. They exhibited highly positive ZP measurements of $-26.3 \pm 4.6$ mV and $-20.07 \pm 0.5$ mV, confirming their high stability in the colloidal solution. Furthermore, the AuNPs and AgNPs had PDI measurements of 0.209 and 0.564, respectively, indicating that AgNPs have a comparatively broader size distribution. In the NaCl study of the AuNPs and AgNPs, the NaCl concentration at which the onset of nanoparticle aggregation occurred was determined to be 150 mm and 80 mM, respectively. This was demonstrated by changes in the AuNP and AgNP colour from red to pink-purple to blue and bright yellow to pale yellow to grey, respectively, indicating the aggregation of nanoparticles. The broadening of SPR peaks plus a decrease in peak intensities, change in ODs, increase in nanoparticle size, lowly negative ZP values, and high PDI confirmed the aggregation behaviour of the nanoparticles. In the $K_2CO_3$ study, successful Au and Ag nanoparticle–antibody conjugation was achieved at 1.2 mM $K_2CO_3$ concentration (pH $\approx$ 7). In the final study, it was found that 4 µg/mL and 6 µg/mL are the optimum antibody concentrations for achieving maximum Au and Ag nanoparticle–antibodies conjugation, respectively, which can help to improve the sensitivity, reproducibility, and limit detection of the lateral flow immunochromatographic assays.

**Author Contributions:** Conceptualization, B.J.; methodology, T.S. and B.J.; software, T.S, V.S. and B.J.; validation, B.J. and F.T.; formal analysis, T.S.; investigation, B.J.; resources, B.J. and F.T.; data curation, B.J.; writing—original draft preparation, T.S. and V.S.; writing—review and editing, B.J. and F.T.; visualization, B.J.; supervision, B.J. and F.T.; project administration, B.J. and F.T.; funding acquisition, B.J. and F.T. All authors have read and agreed to the published version of the manuscript.

**Funding:** This research was funded by the Government of Ireland's postgraduate scholarship programme and from the European Union's Horizon 2020 research and innovation programme under the Marie Skłodowska-Curie grant agreement No. 847402.

**Informed Consent Statement:** Not applicable.

**Data Availability Statement:** The data presented in this study are available upon request from the corresponding author.

**Conflicts of Interest:** The authors declare no conflict of interest.

# References

1. Conrad, M.; Proll, G.; Builes-Münden, E.; Dietzel, A.; Wagner, S.; Gauglitz, G. Tools to Compare Antibody Gold Nanoparticle Conjugates for a Small Molecule Immunoassay. *Microchim. Acta* **2023**, *190*, 62. [CrossRef] [PubMed]
2. Koczula, K.M.; Gallotta, A. Lateral Flow Assays. *Essays Biochem.* **2016**, *60*, 111. [CrossRef]
3. Zhang, L.; Mazouzi, Y.; Salmain, M.; Liedberg, B.; Boujday, S. Antibody-Gold Nanoparticle Bioconjugates for Biosensors: Synthesis, Characterization and Selected Applications. *Biosens. Bioelectron.* **2020**, *165*, 112370. [CrossRef]
4. Razo, S.C.; Elovenkova, A.I.; Safenkova, I.V.; Drenova, N.V.; Varitsev, Y.A.; Zherdev, A.V.; Dzantiev, B.B. Comparative Study of Four Coloured Nanoparticle Labels in Lateral Flow Immunoassay. *Nanomaterials* **2021**, *11*, 3277. [CrossRef]
5. Oliveira, J.P.; Prado, A.R.; Keijok, W.J.; Antunes, P.W.P.; Yapuchura, E.R.; Guimarães, M.C.C. Impact of Conjugation Strategies for Targeting of Antibodies in Gold Nanoparticles for Ultrasensitive Detection of 17β-Estradiol. *Sci. Rep.* **2019**, *9*, 13850. [CrossRef]
6. Jazayeri, M.H.; Amani, H.; Pourfatollah, A.A.; Pazoki-Toroudi, H.; Sedighimoghaddam, B. Various Methods of Gold Nanoparticles (GNPs) Conjugation to Antibodies. *Sens. Biosens. Res.* **2016**, *9*, 17–22. [CrossRef]
7. Ruiz, G.; Tripathi, K.; Okyem, S.; Driskell, J.D. PH Impacts the Orientation of Antibody Adsorbed onto Gold Nanoparticles. *Bioconjug. Chem.* **2019**, *30*, 1182–1191. [CrossRef] [PubMed]
8. Horisberger, M.; Rosset, J.; Bauer, H. Colloidal Gold Granules as Markers for Cell Surface Receptors in the Scanning Electron Microscope. *Experientia* **1975**, *31*, 1147–1149. [CrossRef] [PubMed]
9. Petrakova, A.V.; Urusov, A.E.; Zherdev, A.V.; Dzantiev, B.B. Gold Nanoparticles of Different Shape for Bicolor Lateral Flow Test. *Anal. Biochem.* **2019**, *568*, 7–13. [CrossRef]
10. Zangheri, M.; Di Nardo, F.; Anfossi, L.; Giovannoli, C.; Baggiani, C.; Roda, A.; Mirasoli, M. A Multiplex Chemiluminescent Biosensor for Type B-Fumonisins and Aflatoxin B1 Quantitative Detection in Maize Flour. *Analyst* **2015**, *140*, 358–365. [CrossRef]
11. Keene, A.M.; Tyner, K.M. Analytical Characterization of Gold Nanoparticle Primary Particles, Aggregates, Agglomerates, and Agglomerated Aggregates. *J. Nanopart. Res.* **2011**, *13*, 3465–3481. [CrossRef]
12. Kasoju, A.; Shahdeo, D.; Khan, A.A.; Shrikrishna, N.S.; Mahari, S.; Alanazi, A.M.; Bhat, M.A.; Giri, J.; Gandhi, S. Fabrication of Microfluidic Device for Aflatoxin M1 Detection in Milk Samples with Specific Aptamers. *Sci. Rep.* **2020**, *10*, 4627. [CrossRef]
13. Mirica, A.C.; Stan, D.; Chelcea, I.C.; Mihailescu, C.M.; Ofiteru, A.; Bocancia-Mateescu, L.A. Latest Trends in Lateral Flow Immunoassay (LFIA) Detection Labels and Conjugation Process. *Front. Bioeng. Biotechnol.* **2022**, *10*, 922772. [CrossRef] [PubMed]
14. Sahoo, P.R.; Singh, P. Size Tunable Gold Nanoparticle and Its Characterization for Labeling Application in Animal Health. *Vet. World* **2014**, *7*, 1010–1013. [CrossRef]
15. Wang, Q.; Li, S.; Zhang, Y.; Wang, S.; Guo, J.; Wang, J. A Highly Sensitive Photothermal Immunochromatographic Sensor for Detection of Aflatoxin B1 Based on $Cu_{2-x}Se$-Au Nanoparticles. *Food Chem.* **2023**, *401*, 134065. [CrossRef]
16. Turkevich, J.; Stevenson, P.C.; Hillier, J. A Study of the Nucleation and Growth Processes in the Synthesis of Colloidal Gold. *Discuss. Faraday Soc.* **1951**, *11*, 55–75. [CrossRef]
17. Rashid, M.U.; Bhuiyan, M.K.H.; Quayum, M.E. Synthesis of Silver Nano Particles (Ag-NPs) and Their Uses for Quantitative Analysis of Vitamin C Tablets. *Dhaka Univ. J. Pharm. Sci.* **2013**, *12*, 29–33. [CrossRef]
18. Tam, J.O.; de Puig, H.; Yen, C.W.; Bosch, I.; Gómez-Márquez, J.; Clavet, C.; Hamad-Schifferli, K.; Gehrke, L. A Comparison of Nanoparticle-Antibody Conjugation Strategies in Sandwich Immunoassays. *J. Immunoass. Immunochem.* **2017**, *38*, 355–377. [CrossRef] [PubMed]
19. Lou, S.; Ye, J.Y.; Li, K.Q.; Wu, A. A Gold Nanoparticle-Based Immunochromatographic Assay: The Influence of Nanoparticulate Size. *Analyst* **2012**, *137*, 1174–1181. [CrossRef] [PubMed]
20. Chen, Y.; Xianyu, Y.; Jiang, X. Surface Modification of Gold Nanoparticles with Small Molecules for Biochemical Analysis. *Acc. Chem. Res.* **2017**, *50*, 310–319. [CrossRef]
21. Park, J.; Chariou, P.L.; Steinmetz, N.F. Site-Specific Antibody Conjugation Strategy to Functionalize Virus-Based Nanoparticles. *Bioconjug. Chem.* **2020**, *31*, 1408–1416. [CrossRef] [PubMed]

22. Maddahfar, M.; Wen, S.; Hosseinpour Mashkani, S.M.; Zhang, L.; Shimoni, O.; Stenzel, M.; Zhou, J.; Fazekas De St Groth, B.; Jin, D. Stable and Highly Efficient Antibody-Nanoparticles Conjugation. *Bioconjug. Chem.* **2021**, *32*, 1146–1155. [CrossRef] [PubMed]
23. Khlebtsov, B.N.; Tumskiy, R.S.; Burov, A.M.; Pylaev, T.E.; Khlebtsov, N.G. Quantifying the Numbers of Gold Nanoparticles in the Test Zone of Lateral Flow Immunoassay Strips. *ACS Appl. Nano Mater.* **2019**, *2*, 5020–5028. [CrossRef]
24. Lata, K.; Jaiswal, A.K.; Naik, L.; Sharma, R. Gold Nanoparticles: Preparation, Characterization and Its Stability in Buffer. *Nano Trends J. Nanotechnol. Its Appl.* **2014**, *17*, 1–10.
25. Mudalige, T.; Qu, H.; Van Haute, D.; Ansar, S.M.; Paredes, A.; Ingle, T. Characterization of Nanomaterials: Tools and Challenges. In *Nanomaterials for Food Applications*; Elsevier: Amsterdam, The Netherlands, 2019; pp. 313–353. [CrossRef]
26. Grüttner, C.; Müller, K.; Teller, J.; Westphal, F.; Foreman, A.; Ivkov, R. Synthesis and Antibody Conjugation of Magnetic Nanoparticles with Improved Specific Power Absorption Rates for Alternating Magnetic Field Cancer Therapy. *J. Magn. Magn. Mater.* **2007**, *311*, 181–186. [CrossRef]
27. Di Pasqua, A.J.; Mishler, R.E.; Ship, Y.L.; Dabrowiak, J.C.; Asefa, T. Preparation of Antibody-Conjugated Gold Nanoparticles. *Mater. Lett.* **2009**, *63*, 1876–1879. [CrossRef]
28. Jiang, H.; Su, H.; Wu, K.; Dong, Z.; Li, X.; Nie, L.; Leng, Y.; Xiong, Y. Multiplexed Lateral Flow Immunoassay Based on Inner Filter Effect for Mycotoxin Detection in Maize. *Sens Actuators B Chem* **2023**, *374*. [CrossRef]
29. Greene, M.K.; Richards, D.A.; Nogueira, J.C.F.; Campbell, K.; Smyth, P.; Fernández, M.; Scott, C.J.; Chudasama, V. Forming Next-Generation Antibody-Nanoparticle Conjugates through the Oriented Installation of Non-Engineered Antibody Fragments. *Chem. Sci.* **2017**, *9*, 79–87. [CrossRef]
30. Thorek, D.L.J.; Elias, D.R.; Tsourkas, A. Comparative Analysis of Nanoparticle-Antibody Conjugations: Carbodiimide versus Click Chemistry. *Mol. Imaging* **2009**, *8*, 221–229. [CrossRef]
31. Bélteky, P.; Rónavári, A.; Zakupszky, D.; Boka, E.; Igaz, N.; Szerencsés, B.; Pfeiffer, I.; Vágvölgyi, C.; Kiricsi, M.; Kónya, Z. Are Smaller Nanoparticles Always Better? Understanding the Biological Effect of Size-Dependent Silver Nanoparticle Aggregation Under Biorelevant Conditions. *Int. J. Nanomed.* **2021**, *16*, 3021. [CrossRef]
32. Garcia, M.A. Surface Plasmons in Metallic Nanoparticles: Fundamentals and Applications. *J. Phys. D Appl. Phys.* **2011**, *44*, 283001. [CrossRef]

**Disclaimer/Publisher's Note:** The statements, opinions and data contained in all publications are solely those of the individual author(s) and contributor(s) and not of MDPI and/or the editor(s). MDPI and/or the editor(s) disclaim responsibility for any injury to people or property resulting from any ideas, methods, instructions or products referred to in the content.

Article

# Ultrasound-Assisted Anthocyanins Extraction from Pigmented Corn: Optimization Using Response Surface Methodology

Annisa Nurkhasanah [1], Titouan Fardad [2], Ceferino Carrera [3], Widiastuti Setyaningsih [1,*] and Miguel Palma [3]

1. Department of Food and Agricultural Product Technology, Faculty of Agricultural Technology, Gadjah Mada University, Jalan Flora, Bulaksumur, Depok, Sleman, Yogyakarta 55281, Indonesia; annisa.nurkhasanah@mail.ugm.ac.id
2. Department of Physical Measurements, Institute of Technology of Lannion, CEDEX, 22302 Lannion, France; titouanfardad@gmail.com
3. Department of Analytical Chemistry, Faculty of Sciences, Instituto de Investigación Vitivinícola y Agroalimentaria (IVAGRO), Agrifood Campus of International Excellence (CeiA3), University of Cadiz, Puerto Real, 11510 Cadiz, Spain; ceferino.carrera@uca.es (C.C.); miguel.palma@uca.es (M.P.)
* Correspondence: widiastuti.setyaningsih@ugm.ac.id; Tel.: +62-274-549650

**Citation:** Nurkhasanah, A.; Fardad, T.; Carrera, C.; Setyaningsih, W.; Palma, M. Ultrasound-Assisted Anthocyanins Extraction from Pigmented Corn: Optimization Using Response Surface Methodology. *Methods Protoc.* **2023**, *6*, 69. https://doi.org/10.3390/mps6040069

Academic Editors: Verónica Pino, Victoria Samanidou and Natasa Kalogiouri

Received: 20 June 2023
Revised: 27 July 2023
Accepted: 28 July 2023
Published: 30 July 2023

**Copyright:** © 2023 by the authors. Licensee MDPI, Basel, Switzerland. This article is an open access article distributed under the terms and conditions of the Creative Commons Attribution (CC BY) license (https://creativecommons.org/licenses/by/4.0/).

**Abstract:** This study aimed to determine the optimal UAE conditions for extracting anthocyanins from pigmented corn using the Box–Behnken design (BBD). Six anthocyanins were identified in the samples and were used as response variables to evaluate the effects of the following working variables: extraction solvent pH (2–7), temperature (10–70 °C), solvent composition (0–50% methanol in water), and ultrasound power (20–80%). The extraction time (5–25 min) was evaluated for complete recovery. Response surface methodology suggested optimal conditions, specifically 36% methanol in water with pH 7 at 70 °C using 73% ultrasound power for 10 min. The method was validated with a high level of accuracy (>90% of recovery) and high precision (CV < 5% for both repeatability and intermediate precision). Finally, the proposed analytical extraction method was successfully applied to determine anthocyanins that covered a wide concentration range (36.47–551.92 mg kg$^{-1}$) in several pigmented corn samples revealing potential varieties providing more health benefits.

**Keywords:** anthocyanins; Box–Behnken design; optimization; purple corn; ultrasound-assisted extraction

## 1. Introduction

Pigmented corn is recognized by orange, red, purple, and blue kernels [1,2]. In Indonesia, varieties of purple and rainbow kernels are commonly available. Therefore, anthocyanins are traditionally considered natural food coloring agents [3,4] and have been used in Indonesia to reduce the use of artificial colorants. The varied colors of pigmented corn kernels are defined by anthocyanins, which provide essential health benefits such as anti-oxidant, anti-diabetic, anti-cancer, anti-inflammatory, and anti-obesity [5,6].

Several studies have reported high levels of anthocyanins in pigmented corn. Anthocyanin content in pigmented corn from Australia ranges from 2.2 to 4.4 g kg$^{-1}$ [7]. The highest reported anthocyanin content in whole fresh purple corn (16.4 g kg$^{-1}$) was higher than that in blueberries (3.9 g kg$^{-1}$) [8]. Because it is a prominent source of anthocyanins, a reliable method for identifying and quantifying anthocyanins in various pigmented corn is necessary.

Solid–liquid maceration is the foremost extraction treatment for anthocyanins [9]. In some cases, maceration to extract anthocyanins from purple corn requires a long extraction time of up to 3 h, thereby promoting the degradation of anthocyanins [10–13]. Therefore, advanced technology to accelerate extraction is continuously being developed to increase efficiency [14].

Ultrasound-assisted extraction (UAE) is widely used because it is not limited by the type of solvent, has low solvent consumption, and has a fast extraction time, thereby preventing component damage [15–17]. UAE provides higher anthocyanin recovery than microwave-assisted extraction and maceration [12,18–21] by performing extraction in a shorter time to avoid the breakdown of anthocyanin during the process [14].

The principle of analyte extraction from a solid matrix into a solvent is related to the cavitation effect produced by ultrasound. The type of solvent and composition of the mixture are essential factors because they are necessary for the cavitation effect on the surface of the solid sample. Another critical factor is the solubility of the target compound in the solvent medium. Anthocyanins are highly soluble in water and polar organic solvents, whereas their glycoside forms are very soluble in pure water [3,22]. Therefore, a mixture of polar organic solvents and water is suitable for extracting anthocyanins [23,24]. Methanol was chosen as the extraction solvent because of its higher effectiveness in anthocyanin extraction than ethanol [3,18,25]. A mixture of methanol and water was used to extract anthocyanins from corn [7,26,27].

Factors associated with the chemical properties of the solvent, such as temperature and pH, have been reported to affect extraction recovery [11,28] significantly. Anthocyanins in pigmented corn include cyanidin, pelargonidin, and peonidin [7,27,29], which are generally stable in acidic solutions, but the latter compound is also durable in high-pH solutions [3]. Other factors related to mass transfer effects, such as ultrasound power, pulse duty cycle, and solid: liquid ratio, also considerably affect extraction efficiency [28,30–32].

As many factors must be optimized, a Box–Behnken design (BBD) in conjunction with response surface methodology can help determine the condition providing the highest recovery of anthocyanins from pigmented corn. Therefore, this study aimed to optimize and validate the UAE method for recovering anthocyanins from pigmented corn. The proposed analytical method was subsequently applied to determine anthocyanins in various pigmented corns.

## 2. Materials and Methods

### 2.1. Chemicals and Reagents

HPLC-grade methanol was obtained from Fisher Scientific (Loughborough, UK). The water was purified using a Milli-Q water purification system (Millipore, Bedford, MA, USA). The pH of the extraction solvent was adjusted using 0.1 N hydrochloric acid (HCl), 0.1 M sodium hydroxide (NaOH), and formic acid (Panreac, Barcelona, Spain). Analytical grade cyanidin 3-*O*-glucoside, pelargonidin-3-*O*-glucoside, and peonidin-3-*O*-glucoside were obtained from Sigma-Aldrich (St. Louis, MO, USA).

### 2.2. Samples

Six corn samples, including red, yellow, purple, and mixed-color kernels, were obtained from the local market in Indonesia (Figure 1). The samples were traditionally sun-dried by a farmer and stored at ambient temperature. The dried samples were then milled using a grinder (ML 130 Type SP-7406, JATA, Tudela, Spain) for 5 min with an on-off interval every 30 s and passed through a 1 mm screen mesh using a vibratory sieve shaker (AS 200, Retsch GmbH, Haan, Germany). The sample powder was homogenized and stored in a closed container until analysis. A mixture of samples in the same proportions was used to optimize the UAE experiments.

**Figure 1.** Corn samples collected from Indonesian Provinces: (a) Riau, (b) Lampung, (c) West Java, and (d) Central Java.

*2.3. Ultrasound-Assisted Extraction (UAE)*

Anthocyanin extraction from pigmented corn was performed based on Gonzales et al. [33], with slight modifications. UAE utilized a Sonopuls HD 4200 Ultrasonic Homogenizer (20 Hz, 200 W, Bandelin Electronic GmbH & Co. KG, Heinrichstrabe, Berlin, Germany) with a titanium probe TS 104 (diameter 4.5 mm). The extraction temperature was controlled using a water bath (Frigiterm-10, J.P. Selecta, Barcelona, Spain). The extraction for every 1 g sample employed 25 mL of solvent with a certain percentage of methanol in water at the defined pH based on the experimental design. The pH of the extraction solvent was adjusted using HCl 0.1 N and NaOH 0.1 M and measured using a Crison GLP22 pH meter (Barcelona, Spain). Extraction was carried out according to the experimental design for 5 min using $0.5 \text{ s}^{-1}$ of a pulse duty cycle. Subsequently, the supernatant was separated from the solid material using a centrifuge (Centrofriger-BLT 230V, Selecta, Barcelona, Spain) at 4000 rpm for 10 min. The supernatant was then placed into 25 mL volumetric flasks to adjust the volume and pH (2). A 0.22 µm nylon syringe filter (Membrane Solutions, Dallas, TX, USA) was used to remove impurities before injecting the extract into the UHPLC-UV-Vis system.

*2.4. Determination of the Anthocyanins by UHPLC-UV-Vis*

Anthocyanin content was determined using UHPLC-UV-Vis based on Gonzales et al. [33]. An Elite UHPLC LaChrom System (Hitachi, Tokyo, Japan) equipped with an L-2200U autosampler, an L2300 column oven, and two L-2160U pumps was used for the chromatographic analyses. Separations were performed on a reverse-phase C18 column (2.1 × 50 mm and 2.6 µm particle size; Phenomenex, Kinetex, CoreShell Technology, Torrance, CA, USA). The injection volume was 15 µL. Elution was performed using mobile phases A (water with 5% formic acid) and B (pure methanol). Gradient separation was performed as follows: 2% B, 0.00 min; 2% B, 1.50 min; 15% B, 3.30 min; 15% B, 4.80 min; 35% B, 5.40 min; and 100% B, 6 min. The flow rate was 0.7 mL min$^{-1}$. The column temperature was set to 50 °C. The system was coupled to a UV-Vis detector (L-2420U) and set at 520 nm for anthocyanin quantification. Assuming that the absorbance values of the various anthocyanins were similar and considering their molecular weights, the total anthocyanin content was calculated by summing the detected anthocyanins. The sum of anthocyanin-detected areas was measured and normalized to the experimental design response. A calibration curve prepared based on cyanidin 3-glucoside was used to quantify anthocyanins because more than 70% of the anthocyanins in pigmented corn are cyanidin-based compounds [7]. The analyses were performed in triplicate, and the results were expressed as mg of cyanidin 3-glucoside equivalents (CGE) kg$^{-1}$ of dried corn kernel.

*2.5. Identification of Anthocyanins Using UHPLC-PDA-QToF-MS*

Anthocyanins in mixed samples were identified using a UHPLC system coupled to a photodiode array detector and a quadrupole time-of-flight mass spectrometer (UHPLC-PDA–QToF–MS) model Xevo G2 (Waters Corp., Milford, MA, USA). Identification was based on Gonzales et al. [33]. Separations were performed on a reverse-phase C18 column (100 × 2.1 mm and 1.7 µm particle size). Elution was performed using mobile phases A (water with 2% formic acid) and B (pure methanol). Gradient separation was performed as follows: 5% B, 0.00 min; 20% B, 3.30 min; 30% B, 3.86 min; 40% B, 5.05 min; 55% B, 5.35 min; 60% B, 5.64 min; 95% B, 5.94 min; and 95% B, 7.50 min, with a flow rate of 0.4 mL min$^{-1}$. Each analysis was conducted within 12 min, including 4 min to restore initial conditions. The electrospray was operated in the positive ionization mode. The desolvation gas temperature was 500 °C, and the flow rate was 700 L h$^{-1}$. The source temperature was 150 °C, and the capillary cone was set to 700 V. The cone voltage was set to 20 V with a gas flow of 10 L h$^{-1}$. The trap collision energy was 4 eV. The full scan mode was used to identify anthocyanins in the 100–1200 $m/z$ range.

Six anthocyanins were identified in kernel corn samples (Figure 2). The major anthocyanins [M$^+$] were identified as cyanidin-3-glucoside ($m/z$ 449), pelargonidin-3-glucoside ($m/z$ 493), peonidin-3-glucoside ($m/z$ 463), cyanidin-3-malonyl glucoside ($m/z$ 535), pelargonidin-3-malonyl glucoside ($m/z$ 519), and peonidin-3-malonyl glucoside ($m/z$ 549). The identified anthocyanins were the same as those described previously by Colombo et al. [1].

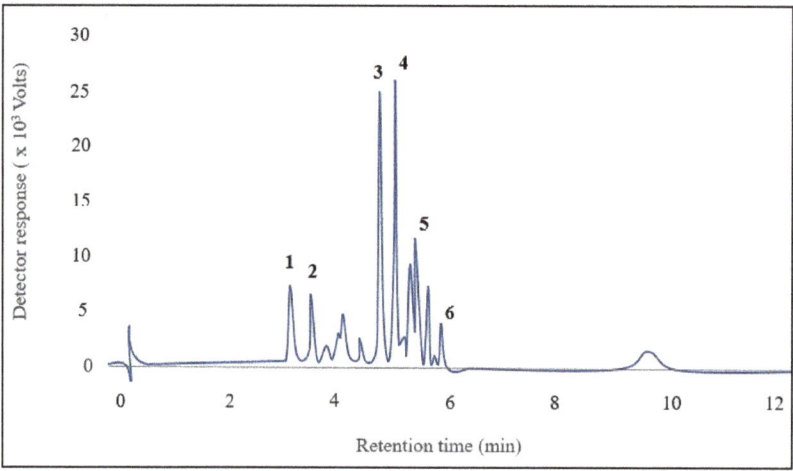

**Figure 2.** Chromatogram of 6 anthocyanins identified in the mixture of pigmented corns: 1. Cyanidin-3-glucoside, 2. Pelargonidin-3-glucoside, 3. Peonidin-3-glucoside, 4. Cyanidin-3-malonyl glucoside, 5. Pelargonidin-3-malonyl glucoside, 6. Peonidin-3-malonyl glucoside.

*2.6. Experimental Design and Statistical Analysis*

Box–Behnken design (BBD) with Response Surface Methodology was used to study factors that may affect the extraction efficiency of anthocyanins from pigmented corn, namely pH, temperature, solvent composition, and sonication power. The total area covered by the six identified anthocyanins was used as the target response. A BBD with four independent variables at three levels of factor values (−1, 0, and 1) was carried out. The independent variables and their levels are listed in Table 1. The overall design consists of 27 basic experimental units, as detailed in Table 2.

Table 1. Selected independent variables and their levels.

| Independent Variables | Levels | | |
| --- | --- | --- | --- |
| | −1 | 0 | +1 |
| $x_1$, pH | 2 | 4.5 | 7 |
| $x_2$, Temperature (°C) | 10 | 40 | 70 |
| $x_3$, Solvent composition (% methanol in water) | 0 | 25 | 50 |
| $x_4$, Ultrasound power (%) | 20 | 50 | 80 |

Table 2. A Box–Behnken design for four factors with the measured responses.

| DOE | Independent Variables | | | | Relative Measured Value to Maximum Responses * (%) |
| --- | --- | --- | --- | --- | --- |
| | $x_1$ | $x_2$ | $x_3$ | $x_4$ | |
| 1 | −1 | 0 | −1 | 0 | 34.55 |
| 2 | −1 | 0 | +1 | 0 | 59.63 |
| 3 | 0 | −1 | −1 | 0 | 35.45 |
| 4 | +1 | +1 | 0 | 0 | 82.22 |
| 5 | 0 | 0 | +1 | +1 | 60.95 |
| 6 | 0 | 0 | −1 | −1 | 29.90 |
| 7 | 0 | 0 | +1 | −1 | 64.09 |
| 8 | 0 | −1 | 0 | −1 | 58.85 |
| 9 | 0 | 0 | −1 | +1 | 42.26 |
| 10 | +1 | 0 | 0 | +1 | 83.42 |
| 11 | 0 | +1 | 0 | +1 | 91.01 |
| 12 | 0 | 0 | 0 | 0 | 74.21 |
| 13 | 0 | −1 | +1 | 0 | 100.00 |
| 14 | −1 | −1 | 0 | 0 | 67.62 |
| 15 | 0 | +1 | +1 | 0 | 83.69 |
| 16 | −1 | 0 | 0 | +1 | 80.78 |
| 17 | 0 | +1 | 0 | −1 | 77.10 |
| 18 | −1 | +1 | 0 | 0 | 75.68 |
| 19 | 0 | 0 | 0 | 0 | 67.73 |
| 20 | 0 | −1 | 0 | +1 | 65.36 |
| 21 | +1 | −1 | 0 | 0 | 80.27 |
| 22 | −1 | 0 | 0 | −1 | 62.75 |
| 23 | +1 | 0 | +1 | 0 | 91.86 |
| 24 | 0 | +1 | −1 | 0 | 40.66 |
| 25 | +1 | 0 | 0 | −1 | 59.63 |
| 26 | 0 | 0 | 0 | 0 | 73.13 |
| 27 | +1 | 0 | −1 | 0 | 57.73 |

* Relative value to the total chromatographic area (520 nm) in the experimental design.

MINITAB (version X) (Minitab LLC, State College, PA, USA) generated the BBD and established the RSM model. An analysis of variance (ANOVA, $p = 0.05$) in conjunction with the Least Significant Difference (LSD, $p = 0.05$) test was used to determine the significance of the difference between the means.

*2.7. Determination of the Optimal Extraction Time*

The optimum extraction condition proposed by the RSM was used to evaluate the extraction time. The optimal extraction time was determined by assessing the level of anthocyanins at different extraction times (5, 10, 15, 20, and 25 min). The experiment was performed in triplicate using a randomized block design.

*2.8. Method Validation*

The analytical method was validated based on ICH Q2(R1) guidelines [34]. In addition to evaluating the analytical properties of the chromatographic procedure, the precision and accuracy assessments of the extraction method were also included in the method validation. The precision of the method was evaluated by performing repeatability (intra-day) and intermediate precision (inter-day). Repeatability was assessed by repeating nine analyses from a sample on the same day ($n = 9$), while intermediate precision was evaluated by performing three extractions on three consecutive days ($n = 3 \times 3$). Precisions were expressed as the coefficient of variation (%CV). The trueness of the method was assessed by calculating the UAE recovery (%R), which was determined by comparing the total anthocyanin areas of the samples with and without spiking. The recovery was performed by adding the concentrated sample as a spike solution in the 25–40% range.

### 3. Results and Discussion

*3.1. Performance of the Chromatographic Method*

The total area of the detected anthocyanins was the response used for the optimization step, and the determination method for the compounds was validated for quality assurance. Hence, chromatographic analysis was validated by measuring the linearity, limit of detection (LOD), limit of quantification (LOQ), and precision of the method. A calibration curve ($y = 168.154x - 28.642$) was prepared using the cyanidin 3-glucoside standard (1–48 mg L$^{-1}$) and measured at 520 nm. The resulting coefficient of determination ($R^2$) was 0.9999, as described by Gonzales et al. [33]. The slope and standard deviation from regression were included in the calculation to define LOD (0.38 mg L$^{-1}$) and LOQ (1.18 mg L$^{-1}$). The molecular weights of the other anthocyanins (pelargonidin, peonidin, and malonyl derivatives) were used to calculate the total level of anthocyanins and were expressed as cyanidin glucoside equivalents (CGE). The precision of the peak area in the chromatographic results was evaluated by performing intra-day (CV, 1.52%) and inter-day (CV, 2.11%) injections on the sample. The resulting CV values were less than the acceptable level (2.7%) for analyte concentrations of 1–48 mg L$^{-1}$, confirming the high precision of the method [35].

*3.2. Effect of the UAE Operating Variables in the Recovery of Anthocyanins*

The BBD, consisting of 27 experiments, including three central points, was completed. Subsequently, the effect of the studied UAE factors on the level of anthocyanins extracted from corn kernels was assessed using analysis of variance (ANOVA). The statistical significance of each effect provided by the UAE factors was obtained by comparing the mean square error with the estimated experimental error. The standardized values ($p = 0.05$) in descending order of importance are plotted on a Pareto chart for the main, interaction, and quadratic effects (Figure 3).

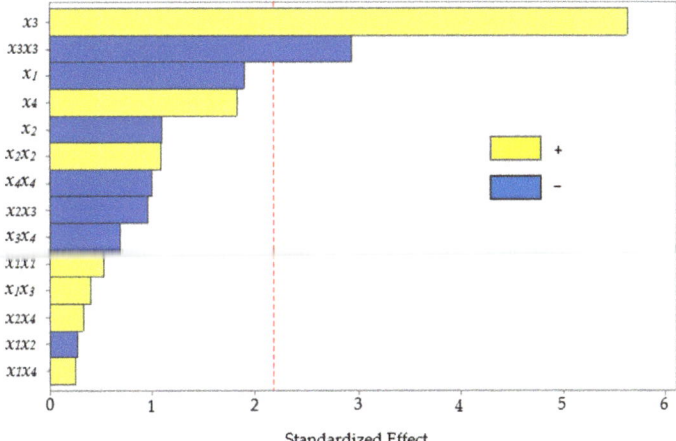

**Figure 3.** Pareto chart of the standardized effects from the UAE factors ($x_1$, pH; $x_2$, temperature; $x_3$, solvent composition; and $x_4$, power of sonication).

The bars crossing the vertical line indicate the factors or combinations that significantly affect the response. The two effects had *p*-values below 0.05, marking a significant difference at the 95% confidence level. The main effect of solvent composition positively influenced ($p < 0.001$) the extraction, which means that the higher the amount of methanol in water, the higher the recovery of anthocyanins. The solvent composition has previously been reported as an influential variable for the extraction of anthocyanins from other similar matrices, such as purple corn cobs [4,36], purple corn flour [37], and red rice bran [38]. Hydroalcoholic mixtures are also more efficient than pure solvents for extracting moderately polar or amphiphilic molecules such as anthocyanins [39,40]. The polarity agreement between the analyte and the extraction solvent could increase the solubility. Improving the methanol in water as the extraction solvent provided higher recovery because the solubility of anthocyanins increased, facilitating the mass transfer rate.

However, the quadratic solvent composition showed a significant effect ($p = 0.013$) with an inverse relationship, which means that excess methanol in the solvent lowered the recovery. Excess methanol in the solvent would decrease the extraction yield because it creates a difference in polarity between the analyte and the extraction solvent, affecting the solubility.

*3.3. Prediction Model Using Response Surface Methodology*

The purpose of optimization using Response Surface Methodology (RSM) was to obtain the best combination of UAE factors to achieve the highest recovery results. A mathematical model of a second-order polynomial Equation (1) was established, considering the significant main and quadratic effects of the solvent composition on the level of extracted anthocyanins.

$$y = 0.7169 + 0.1831 x_3 - 0.1431 x_3 x_3 \tag{1}$$

where $y$ is the total area of anthocyanin as a response, $x_3$ is the solvent composition, and $x_3 x_3$ is the quadratic solvent composition. The correlation between the measured and predicted values of total anthocyanins obtained using the model is plotted in Figure 4.

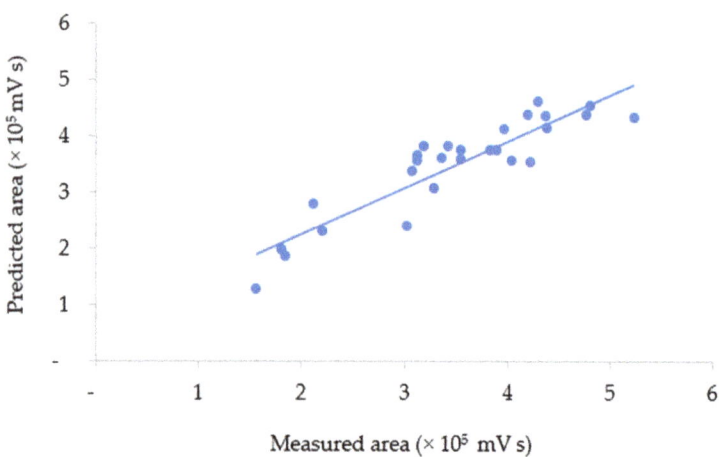

**Figure 4.** The measured and predicted areas of total anthocyanins in UAE extract.

The agreement between the measured and predicted total anthocyanin values was evaluated by the average relative prediction error (9.73%). The measured vs. predicted data values showed low variability (0.03–26.60%) around the mean value. A lack-of-fit test was also performed to determine whether the selected model was suitable for describing the measured data results. The resulting $p$-value for lack of fit (0.07) was non-significant, indicating that the model chosen satisfactorily represented the data at the 95% confidence level [30,31,39]. Hence, the equation model could describe the conditions of the UAE factors that defined the response with satisfactory predictions, as plotted in Figure 5.

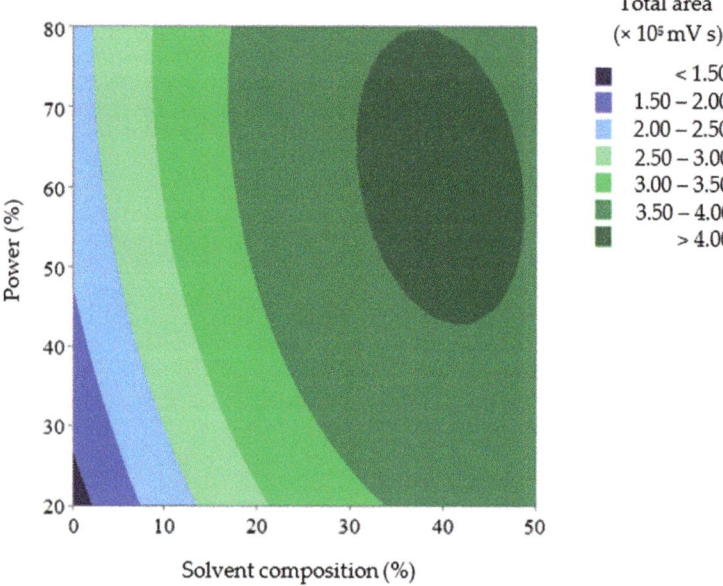

**Figure 5.** 2D Surface plot diagram of the level of anthocyanins in UAE extracts by solvent composition and power.

The dark blue area represents the lowest relative area, whereas the dark green area represents the highest relative area. In this case, the region representing the highest recovery in the DOE was obtained when the solvent composition and ultrasound power approached their highest values in the studied range.

*3.4. Optimization Conditions and Verification*

The experimental design results suggested that the optimal extraction condition was set at +0.43 for solvent composition, while the remaining conditions were at a level of +1 for pH, +1 for temperature, and +0.78 for ultrasound power. Therefore, optimized extraction of anthocyanins from mixed pigmented corn samples using UAE could be achieved by applying 35.86% methanol in water at a pH of 7, a temperature of 70 °C, and a sonication power of 73.33%. The predicted anthocyanin area was 495.52 V s.

Three experiments were conducted to verify the optimum condition using 36% methanol in water with a pH of 7 at 70 °C and a sonication power of 73%. The resulting level of anthocyanins, indicated by the peak area, was 449.57 ± 3.44 V s, with a deviation of 9.27% from the prediction. These findings suggest that the Box–Behnken design successfully optimized anthocyanin extraction with RSM.

Muangrat et al. [4] conducted optimization research on extracting anthocyanin from purple corn cobs using UAE. The optimal conditions were similar to those determined in this study, namely the temperature (64 °C). However, the solvent composition (50% ethanol in water) and ultrasound amplitude (50 °C) differed from those observed in this study; both factors could be closely related to the various parts of corn.

*3.5. Optimal Extraction Time*

The optimal extraction time was determined by varying the extraction time (5, 10, 15, 20, and 25 min). An ANOVA was used to analyze the results, which showed that the extraction time significantly affected the extraction efficiency ($p = 0.018$). As shown in Figure 6, when the extraction time was increased to 10 min, the level of anthocyanins in the extract also increased. However, this level decreased when the extraction time was increased to 20 min. A prolonged heating process may result in the degradation of thermolabile compounds [41]. Extracted anthocyanins are also easily oxidized by the environment if the extraction time is too long at high temperatures [42]. Thus, the extraction time of 10 min was chosen as the optimal extraction time for anthocyanins in pigmented corn. This value was much lower than the extraction times optimized in purple corn cobs with UAE (30 min) [4], MAE (19 min) [13], and in purple corn kernels with stirred maceration (30 min) [1].

*3.6. Validation of the UAE Method*

The accuracy of the method was assessed by measuring UAE recovery (%R). The anthocyanin extraction recovery was 92.81%. According to the AOAC guidelines (80–110%), the recovery results were within the accepted range, indicating that the optimal method has a high level of accuracy [35]. The precision of the method was evaluated by performing repeatability (intra-day) and intermediate precision (inter-day) tests. The results of the repeatability and intermediate precision tests were 4.62 and 4.61%, respectively. The acceptance limit for the precision of the analyte at a concentration of 1–10 mg $L^{-1}$ was 11.0% [35]. The CV value obtained was less than the acceptance limit by the AOAC; thus, the proposed method can be validated as providing high-precision results.

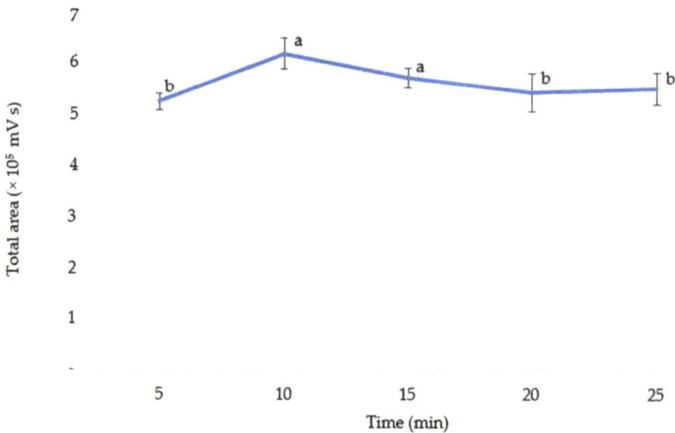

**Figure 6.** The total area of anthocyanins according to the different extraction times. The means with different letters are significantly different ($p = 0.05$).

*3.7. Applying the Optimized Method to Different Pigmented Corn*

The extraction method was applied to extract anthocyanins from pigmented corn to recover anthocyanins from various samples (Table 3). The studied pigmented corn contained anthocyanins in the 36.47–551.92 mg CGE kg$^{-1}$ range. Purple corn had the highest anthocyanin content compared with red and white-purple corn. Purple corn has a high chroma and is positively correlated with anthocyanins [36].

**Table 3.** Anthocyanin content of different pigmented corn matrices.

| Location | Color | Picture | Anthocyanin Content (mg CGE kg$^{-1}$) |
|---|---|---|---|
| Lampung | Red | | 281.56 ± 31.29 |
| West Java | Red | | 36.47 ± 6.65 |
| Riau | Purple | | 551.92 ± 14.02 |
| Central Java | Purple and white | | 47.01 ± 4.32 |

Table 3. *Cont.*

| Location | Color | Picture | Anthocyanin Content (mg CGE kg$^{-1}$) |
|---|---|---|---|
| Central Java | Yellow | | <LOQ |
| Lampung | Yellow | | <LOQ |

Red corn from Lampung had a higher anthocyanin content than the red variety from West Java. This is because red corn from Lampung has thicker red corn bran than that from West Java. In addition, red corn from West Java has a thicker white endosperm than red corn from Lampung. Purple corn from other regions of Indonesia also contains different anthocyanins. For example, Manado has an anthocyanin content of 341 mg CGE L$^{-1}$, and Malang has an anthocyanin content of 376 mg CGE L$^{-1}$ [10]. The anthocyanin content in corn can vary depending on several factors, one of which is the growing conditions, such as growth location [43].

## 4. Conclusions

A new ultrasound-assisted extraction (UAE) method was successfully optimized using the Box–Behnken design to extract anthocyanins from pigmented corn. The optimal condition was achieved by applying a solvent pH of 7 at 70 °C employing 36% methanol in water and 73% sonication power for 10 min. The proposed optimal method was validated with high levels of accuracy and precision. The developed UAE method was successfully applied to determine the anthocyanin content of pigmented corn from several regions of Indonesia. Pigmented corn contains anthocyanins in the 36.47–551.92 mg CGE kg$^{-1}$ range. Based on these results, it can be concluded that the newly developed UAE method in this study is fast yet reliable for determining anthocyanin content in pigmented corn matrices.

**Author Contributions:** Conceptualization, W.S. and M.P.; methodology, W.S. and M.P.; software, A.N. and C.C.; validation, C.C. and M.P.; formal analysis, A.N. and W.S.; investigation, A.N. and T.F.; resources, A.N., T.F., C.C., and M.P.; data curation, W.S. and M.P.; writing—original draft preparation, A.N. and M.P.; writing—review and editing, W.S. and C.C.; visualization, M.P.; supervision, W.S. and M.P.; project administration, A.N. All authors have read and agreed to the published version of the manuscript.

**Funding:** This research received no external funding.

**Institutional Review Board Statement:** Not applicable.

**Informed Consent Statement:** Not applicable.

**Data Availability Statement:** The data presented in this study are available in this article.

**Acknowledgments:** The authors are grateful to the Instituto de Investigación Vitivinícola y Agroalimentaria (IVAGRO) for providing the necessary support to complete the research.

**Conflicts of Interest:** The authors declare no conflict of interest.

## References

1. Colombo, F.; Di Lorenzo, C.; Petroni, K.; Silano, M.; Pilu, R.; Falletta, E.; Biella, S.; Restani, P. Pigmented Corn Varieties as Functional Ingredients for Gluten-Free Products. *Foods* **2021**, *10*, 1770. [CrossRef] [PubMed]
2. Oladzadabbasabadi, N.; Nafchi, A.M.; Ghasemlou, M.; Ariffin, F.; Singh, Z.; Al-Hassan, A. Natural anthocyanins: Sources, extraction, characterization, and suitability for smart packaging. *Food Packag. Shelf Life* **2022**, *33*, 100872. [CrossRef]

3. Khoo, H.E.; Azlan, A.; Tang, S.T.; Lim, S.M. Anthocyanidins and anthocyanins: Colored pigments as food, pharmaceutical ingredients, and the potential health benefits. *Food Nutr. Res.* **2017**, *61*, 1361779. [CrossRef] [PubMed]
4. Muangrat, R.; Pongsirikul, I.; Blanco, P.H. Ultrasound assisted extraction of anthocyanins and total phenolic compounds from dried cob of purple waxy corn using response surface methodology. *J. Food Process. Preserv.* **2017**, *42*, e13447. [CrossRef]
5. Blando, F.; Calabriso, N.; Berland, H.; Maiorano, G.; Gerardi, C.; Carluccio, M.A.; Andersen, Ø.M. Radical Scavenging and Anti-Inflammatory Activities of Representative Anthocyanin Groupings from Pigment-Rich Fruits and Vegetables. *Int. J. Mol. Sci.* **2018**, *19*, 169. [CrossRef] [PubMed]
6. Tan, J.; Han, Y.; Han, B.; Qi, X.; Cai, X.; Ge, S.; Xue, H. Extraction and purification of anthocyanins: A review. *J. Agric. Food Res.* **2022**, *8*, 100306. [CrossRef]
7. Hong, H.T.; Netzel, M.E.; O'Hare, T.J. Optimisation of extraction procedure and development of LC–DAD–MS methodology for anthocyanin analysis in anthocyanin-pigmented corn kernels. *Food Chem.* **2020**, *319*, 126515. [CrossRef]
8. Lieberman, S. The Antioxidant Power of Purple Corn: A Research Review. *Altern. Complement. Ther.* **2007**, *13*, 107–110. [CrossRef]
9. Teng, Z.; Jiang, X.; He, F.; Bai, W. Qualitative and Quantitative Methods to Evaluate Anthocyanins. *Efood* **2020**, *1*, 339–346. [CrossRef]
10. Chayati, I.; Marsono, Y.; Astuti, M. The Effect of Varieties, Sifting Factions, and Solvents on Total Anthocyanins, Total Phenolic Contents, and Antioxidant Activities of Purple Corn Extracts. *J. Riset. Teknol. Ind.* **2020**, *14*, 13–26. [CrossRef]
11. Chen, L.; Yang, M.; Mou, H.; Kong, Q. Ultrasound-assisted extraction and characterization of anthocyanins from purple corn bran. *J. Food Process. Preserv.* **2017**, *42*, e13377. [CrossRef]
12. Fernandez-Aulis, F.; Hernandez-Vazquez, L.; Aguilar-Osorio, G.; Arrieta-Baez, D.; Navarro-Ocana, A. Extraction and Identification of Anthocyanins in Corn Cob and Corn Husk from Cacahuacintle Maize. *J. Food Sci.* **2019**, *84*, 954–962. [CrossRef]
13. Yang, Z.; Zhai, W. Optimization of microwave-assisted extraction of anthocyanins from purple corn (*Zea mays* L.) cob and identification with HPLC–MS. *Innov. Food Sci. Emerg. Technol.* **2010**, *11*, 470–476. [CrossRef]
14. Salacheep, S.; Kasemsiri, P.; Pongsa, U.; Okhawilai, M.; Chindaprasirt, P.; Hiziroglu, S. Optimization of ultrasound-assisted extraction of anthocyanins and bioactive compounds from butterfly pea petals using Taguchi method and Grey relational analysis. *J. Food Sci. Technol.* **2020**, *57*, 3720–3730. [CrossRef]
15. Albero, B.; Tadeo, J.L.; Pérez, R.A. Ultrasound-assisted extraction of organic contaminants. *TrAC Trends Anal. Chem.* **2019**, *118*, 739–750. [CrossRef]
16. Kumar, K.; Srivastv, S.; Sharanagat, V.S. Ultrasound assisted extraction (UAE) of bioactive compounds from fruit and vegetable processing by-products: A review. *Ultrason. Sonochem.* **2020**, *70*, 105325. [CrossRef] [PubMed]
17. Ikhtiarini, A.N.; Setyaningsih, W.; Rafi, M.; Aminah, N.S.; Insanu, M.; Imawati, I.; Rohman, A. Optimization of Ultrasound-Assisted Extraction and the Antioxidant Activities of Sidaguri (*Sida rhombifolia*). *J. Appl. Pharm. Sci.* **2021**, *11*, 70–76. [CrossRef]
18. Aliaño-González, M.J.; Jarillo, J.A.; Carrera, C.; Ferreiro-González, M.; Álvarez, J.; Palma, M.; Ayuso, J.; Barbero, G.F.; Espada-Bellido, E. Optimization of a Novel Method Based on Ultrasound-Assisted Extraction for the Quantification of Anthocyanins and Total Phenolic Compounds in Blueberry Samples (*Vaccinium corymbosum* L.). *Foods* **2020**, *9*, 1763. [CrossRef] [PubMed]
19. Celli, G.B.; Ghanem, A.; Brooks, M.S.-L. Optimization of ultrasound-assisted extraction of anthocyanins from haskap berries (*Lonicera caerulea* L.) using Response Surface Methodology. *Ultrason. Sonochem.* **2015**, *27*, 449–455. [CrossRef] [PubMed]
20. Vázquez-Espinosa, M.; de Peredo, A.V.G.; Ferreiro-González, M.; Carrera, C.; Palma, M.; Barbero, G.F.; Espada-Bellido, E. Assessment of Ultrasound Assisted Extraction as an Alternative Method for the Extraction of Anthocyanins and Total Phenolic Compounds from Maqui Berries (*Aristotelia chilensis* (Mol.) Stuntz). *Agronomy* **2019**, *9*, 148. [CrossRef]
21. Backes, E.; Pereira, C.; Barros, L.; Prieto, M.; Genena, A.K.; Barreiro, M.F.; Ferreira, I.C. Recovery of bioactive anthocyanin pigments from *Ficus carica* L. peel by heat, microwave, and ultrasound based extraction techniques. *Food Res. Int.* **2018**, *113*, 197–209. [CrossRef]
22. Pérez-Gregorio, R.M.; García-Falcón, M.S.; Simal-Gándara, J.; Rodrigues, A.S.; Almeida, D.P. Identification and quantification of flavonoids in traditional cultivars of red and white onions at harvest. *J. Food Compos. Anal.* **2010**, *23*, 592–598. [CrossRef]
23. Aliaño-González, M.J.; Espada-Bellido, E.; González, M.F.; Carrera, C.; Palma, M.; Ayuso, J.; Álvarez, J.; Barbero, G.F. Extraction of Anthocyanins and Total Phenolic Compounds from Açai (*Euterpe oleracea* Mart.) Using an Experimental Design Methodology. Part 2: Ultrasound-Assisted Extraction. *Agronomy* **2020**, *10*, 326. [CrossRef]
24. Kapasakalidis, P.G.; Rastall, R.A.; Gordon, M.H. Extraction of Polyphenols from Processed Black Currant (*Ribes nigrum* L.) Residues. *J. Agric. Food Chem.* **2006**, *54*, 4016–4021. [CrossRef] [PubMed]
25. Trikas, E.D.; Papi, R.M.; Kyriakidis, D.A.; Zachariadis, G.A. A Sensitive LC-MS Method for Anthocyanins and Comparison of Byproducts and Equivalent Wine Content. *Separations* **2016**, *3*, 18. [CrossRef]
26. Kim, H.Y.; Lee, K.Y.; Kim, M.; Hong, M.; Deepa, P.; Kim, S. A Review of the Biological Properties of Purple Corn (*Zea mays* L.). *Sci. Pharm.* **2023**, *91*, 6. [CrossRef]
27. Suriano, S.; Balconi, C.; Valoti, P.; Redaelli, R. Comparison of total polyphenols, profile anthocyanins, color analysis, carotenoids and tocols in pigmented maize. *LWT* **2021**, *144*, 111257. [CrossRef]
28. Xue, H.; Tan, J.; Li, Q.; Tang, J.; Cai, X. Optimization Ultrasound-Assisted Deep Eutectic Solvent Extraction of Anthocyanins from Raspberry Using Response Surface Methodology Coupled with Genetic Algorithm. *Foods* **2020**, *9*, 1409. [CrossRef] [PubMed]
29. Moreno, Y.S.; Sánchez, G.S.; Hernández, D.R.; Lobato, N.R. Characterization of Anthocyanin Extracts from Maize Kernels. *J. Chromatogr. Sci.* **2005**, *43*, 483–487. [CrossRef] [PubMed]

30. Albuquerque, B.R.; Pinela, J.; Barros, L.; Oliveira, M.B.P.; Ferreira, I.C. Anthocyanin-rich extract of jabuticaba epicarp as a natural colorant: Optimization of heat- and ultrasound-assisted extractions and application in a bakery product. *Food Chem.* **2020**, *316*, 126364. [CrossRef]
31. Türker, D.A.; Doğan, M. Ultrasound-assisted natural deep eutectic solvent extraction of anthocyanin from black carrots: Optimization, cytotoxicity, in-vitro bioavailability and stability. *Food Bioprod. Process.* **2022**, *132*, 99–113. [CrossRef]
32. Setyaningsih, W.; Saputro, I.E.; Carrera, C.A.; Palma, M. Optimisation of an ultrasound-assisted extraction method for the simultaneous determination of phenolics in rice grains. *Food Chem.* **2019**, *288*, 221–227. [CrossRef] [PubMed]
33. González, M.J.A.; Carrera, C.; Barbero, G.F.; Palma, M. A comparison study between ultrasound–assisted and enzyme–assisted extraction of anthocyanins from blackcurrant (*Ribes nigrum* L.). *Food Chem. X* **2022**, *13*, 100192. [CrossRef] [PubMed]
34. ICH International Conference on Harmonisation of Technical Requirements for Registration of Pharmaceuticals for Human Use. ICH Harmonised Tripartite Guideline Validation of Analytical Procedures: Text And Methodology Q2(R1) Geneva, Switzerland, 2005. Available online: https://www.gmp-compliance.org/files/guidemgr/Q2(R1).pdf (accessed on 1 January 2023).
35. AOAC International. Guidelines for Standard Method Performance Requirements, Appendix F, 1-18, 2016. Available online: https://www.aoac.org/wp-content/uploads/2019/08/app_f.pdf (accessed on 1 January 2023).
36. Yang, Z.; Fan, G.; Gu, Z.; Han, Y.; Chen, Z. Optimization extraction of anthocyanins from purple corn (*Zea mays* L.) cob using tristimulus colorimetry. *Eur. Food Res. Technol.* **2007**, *227*, 409–415. [CrossRef]
37. Ursu, M.G.S.; Milea, A.; Păcularu-Burada, B.; Dumitrașcu, L.; Râpeanu, G.; Stanciu, S.; Stănciuc, N. Optimizing of the extraction conditions for anthocyanin's from purple corn flour (*Zea mays* L.): Evidences on selected properties of optimized extract. *Food Chem. X* **2023**, *17*, 100521. [CrossRef]
38. Wang, Y.; Zhao, L.; Zhang, R.; Yang, X.; Sun, Y.; Shi, L.; Xue, P. Optimization of ultrasound-assisted extraction by response surface methodology, antioxidant capacity, and tyrosinase inhibitory activity of anthocyanins from red rice bran. *Food Sci. Nutr.* **2020**, *8*, 921–932. [CrossRef] [PubMed]
39. Espada-Bellido, E.; Ferreiro-González, M.; Carrera, C.; Palma, M.; Álvarez, J.A.; Barbero, G.F.; Ayuso, J. Extraction of Antioxidants from Blackberry (*Rubus ulmifolius* L.): Comparison between Ultrasound- and Microwave-Assisted Extraction Techniques. *Agronomy* **2019**, *9*, 745. [CrossRef]
40. López, C.J.; Caleja, C.; Prieto, M.; Barreiro, M.F.; Barros, L.; Ferreira, I.C. Optimization and comparison of heat and ultrasound assisted extraction techniques to obtain anthocyanin compounds from Arbutus unedo L. Fruits. *Food Chem.* **2018**, *264*, 81–91. [CrossRef]
41. Xu, D.P.; Li, Y.; Meng, X.; Zhou, T.; Zhou, Y.; Zheng, J.; Zhang, J.J.; Li, H. Bin Natural Antioxidants in Foods and Medicinal Plants: Extraction, Assessment, and Resources. *Int. J. Mol. Sci.* **2017**, *18*, 96. [CrossRef] [PubMed]
42. Minh, T.N.; Ngoc, H.D.H.; Quang, M.V.; Van, T.N. Effect of extraction methods and temperature preservation on total anthocyanins compounds of Peristrophe bivalvis L. Merr leaf. *J. Appl. Biol. Biotechnol.* **2022**, *10*, 1–5. [CrossRef]
43. Zhu, F. Anthocyanins in cereals: Composition and health effects. *Food Res. Int.* **2018**, *109*, 232–249. [CrossRef] [PubMed]

**Disclaimer/Publisher's Note:** The statements, opinions and data contained in all publications are solely those of the individual author(s) and contributor(s) and not of MDPI and/or the editor(s). MDPI and/or the editor(s) disclaim responsibility for any injury to people or property resulting from any ideas, methods, instructions or products referred to in the content.

*Protocol*

# Determination of Polycyclic Aromatic Hydrocarbons (PAHs) in Leaf and Bark Samples of *Sambucus nigra* Using High-Performance Liquid Chromatography (HPLC)

Fausto Viteri [1], Nazly E. Sánchez [2] and Katiuska Alexandrino [3,*]

[1] Grupo de Protección Ambiental (GPA), Facultad de Ciencias de la Ingeniería e Industrias, Universidad UTE, Quito 170527, Ecuador
[2] Departamento de Ingeniería Ambiental y Sanitaria, Universidad del Cauca, Popayan 190007, Colombia
[3] Facultad de Ingeniería y Ciencias Aplicadas, Ingeniería Agroindustrial, Universidad de Las Américas, Quito 170503, Ecuador
* Correspondence: katiuska.alexandrino@udla.edu.ec

**Abstract:** Polycyclic aromatic hydrocarbons (PAHs) are ubiquitous organic compounds coming from natural or anthropogenic activities. Tree organs such as leaves and barks have been used to monitor urban air quality and have achieved remarkable ecological importance. However, the potential of many tree species as biomonitors is still unknown and efforts should be focused on conducting studies that analyze their capabilities with a viable analytical method. In this work, an analytical method for quantification of the 16 EPA priority PAHs from the leaves and bark of *Sambucus nigra* was validated. In general, the method showed good linearity, detection limits, precision, and recoveries, demonstrating that it is suitable for analyzing PAHs in both the leaves and bark of the *Sambucus nigra* species for which no analytical method for PAHs is yet available. The high prevalence of fluoranthene in the samples, which is a PAH related to coal combustion and biomass burning, and benzo[a]pyrene, which has a carcinogenic effect, was identified.

**Keywords:** PAH; biomonitoring; air pollution; tree; HPLC

## 1. Introduction

Polycyclic aromatic hydrocarbons (PAHs) are a series of organic compounds containing two or more fused benzene rings that form during the incomplete combustion of organic matter [1,2]. Their emissions may be due to natural or anthropogenic activities [3]. Approximately 500 different PAHs have been detected in the air [4]. However, only 16 PAHs have been classified by the United States Environmental Protection Agency (U.S. EPA) as priority pollutants due to their high carcinogenic and mutagenic potential [5]. These are: naphthalene (Naph), acenaphthylene (Acy), acenaphthene (Ace), fluorene (Fluo), phenanthrene (Phen), anthracene (Ant), fluoranthene (Flt), pyrene (Pyr), benzo[a]anthracene (BaA), chrysene (Chry), benzo[b]fluoranthene (BbF), benzo[k]fluoranthene (BkF), benzo[a]pyrene (BaP), dibenzo[a,h]anthracene (DahA), benzo[g,h,i]perylene (BghiP), and indeno [1,2,3-cd]pyrene (IcdP).

PAHs exist in the atmosphere in the vapor and/or in particle-bound phase, and a large portion of them are scavenged by vegetation via dry and wet deposition [6]. In this sense, the use of vegetation, especially trees, in the assessment of atmospheric PAHs' concentrations has gained great interest due to its low cost. Moreover, due to their high spatial and temporal distribution, the use of trees provides the possibility of building high-resolution maps of air pollution to detect risk areas in urban areas. However, differences in the ability to accumulate PAHs between tree species have been identified [7–9].

The interception of pollutants by trees take place mainly in the upper portion of the tree, such as leaves, stems, and barks. In this sense, different works have addressed the use of leaves/needles from different tree species to assess the presence of PAHs in urban environments [7–9]. Stomata and outer cuticular lamellae are main vias for the uptake of PAHs in the vapor phase, whereas particle-bound PAHs are accumulated on the leaf surface [9,10]. Other vegetative parts of the tree, such as bark, have been less studied, although some works have shown its good capacity to accumulate PAHs due to its high lipid content, and porous and almost inert surface [11–13].

The evaluation of the atmospheric PAH concentrations using the leaves and barks of different tree species is possible due to the development and application, in recent years, of some analytical procedures. The processes proposed in the literature vary, due to the complexity of the sample matrix. However, some steps in those protocols are similar, including sample pre-treatment, extraction, clean-up, pre-concentration, and chemical analysis. Furthermore, the ways of carrying out these processes are diverse.

Considering the sample pre-treatment, some studies include the use of drying techniques such as freeze drying [14], stoves [15], and ovens [16]. Moreover, crushing techniques using mortars [17], high-speed grinders [18], or liquid nitrogen [19] can also be used. However, there are many works where the intact samples are used, without any prior drying or crushing treatment [11,20–22]. The pre-treatment step has been shown to be a bottleneck in achieving adequate recoveries. Hence, it is important to pay attention to how the samples are prepared, as using severe methods can greatly reduce these recoveries [14].

Regarding PAHs extraction, ultrasonic extraction [11,18,20], Soxhlet extraction [18,21,23], accelerated solvent extraction [18,22,24], and microwave-assisted extraction [11,25] are the most used techniques, which involve the use of different organic solvents for better yields.

Extract clean-up, which is a step that is often necessary to remove some matrix co-extractant compounds, such as lipidic compounds and chlorophylls, which could cause interference and introduce errors in the analysis [26,27], is usually performed by column chromatography or solid-phase extraction (SPE) cartridges with different sorbents such as florisil [7,28], silica gel [13,21,29], or alumina [30,31]. Regarding pre-concentration, the rotary evaporator [17] and the nitrogen stream [32] are the common techniques used. For instrumental analysis, gas chromatography coupled to mass spectrometry (GC-MS) is the most widely used equipment for the detection and quantification of PAHs [14,22–25,33]. High-performance liquid chromatography with diode array (HPLC-DAD) and/or fluorescence detectors (HPLC-Fl) is another technique used, although to a lesser extent than GC-MS [20,21,32].

Although several analytical methods for the identification and quantification of PAHs have been developed in recent years, a method developed for one tree species may not be suitable for other ones. It may not even be suitable for another vegetative part of the same tree. Therefore, the development of accurate and sensitive analytical methods is necessary for the determination of PAHs in different tree species and their vegetative parts. To the best of our knowledge, no study has been carried out for the leaves and bark of *Sambucus nigra*. This is a deciduous multi-stemmed small tree native to Europe, southwestern Asia, and northern Africa, and introduced and widely dispersed in Ecuador and South America in general. This tree species has different medicinal and food uses. On the one hand, their flowers and fruits have flavonoids, organic acids, essential oils, phenolic acids, and anthocyanins showing an antiviral effect, strengthening the immune system and providing inmuno-protection [34]. Moreover, their leaves, berries, and flowers seem to act as antioxidants by neutralizing free radicals [35]. On the other hand, the fruit provides flavor and color to certain foods, and it is used to prepare preserves, wines [36], sponge cakes [37], among other foods.

*Sambucus nigra* can be found in pedestrian areas, parks, and main streets in urban and sub-urban areas, being useful for extensive spatio–temporal sampling; thus, its study as a biomonitor is interesting and necessary. Therefore, the aim of this work was to present an analytical method for the quantitative extraction and determination of 16 US-EPA PAHs in leaf and bark samples of *Sambucus nigra*. The analytical procedure includes the use of an ultrasonic bath for extraction, concentration, an SPE clean-up procedure, and the final concentration before analysis by high-performance liquid chromatography (HPLC).

## 2. Materials and Methods

### 2.1. Sample Collection

Leaf and bark samples of *Sambucus nigra* were collected in a residential area (0°10′38.5″ S 78°21′51.1″ W) in the city of Quito, Ecuador. Sampling was performed from all directions of the tree at a specific height and using a new pair of powder-free vinyl gloves for each sample to avoid cross contamination. Specifically, eight branches were collected from the outer part of the tree by using a pruning shear and at a height of approximately 2 m above the ground. On the other hand, the bark was carefully removed from the boles of the tree at a height of approximately 1.5 m above the ground using a steel knife. Between each sampling, the pruning shear and the steel knife were cleaned with alcohol.

All the collected branches and barks were packed together in a single Ziplock bag, respectively. The bags were then labeled on site with the name of the species tree, sample type (leaf or bark), the date of collection, and GPS coordinates. To avoid photochemical degradation and volatilization of PAHs, bags were wrapped in aluminum to protect them from light and placed in cooler containing ice packs. Finally, samples were transported to the laboratory and stored at −20 °C for four days, for subsequent sample treatment and chemical analysis.

### 2.2. Sample Treatment

Prior to extraction, the samples were defrosted in a desiccator. Then, 2 g of leaves, of identical length and with no evidence of chlorosis or necrosis, were randomly taken by hand from the branches, taking care to minimize contact with the leaf surface, and weighed in six 250 mL beakers. Likewise, 2 g of bark, without the presence of mold, fungi, lichens, or foreign material such as spider webs, was weighed in six 250 mL beakers. Powder-free vinyl gloves were used to avoid cross contamination (a new pair between each weighing).

To evaluate the performance of the method (%Recovery (%R)), 0.3 mL of a 10 µg mL$^{-1}$ certified standard mixture of 16 EPA PAHs in acetonitrile (SigmaAldrich, purchased from Supelco, Ecuador) was added into three of the six beakers with leaves and barks, respectively (final concentration of 1.5 µg g$^{-1}$). This allows determining the recoveries by the matrix spike method in triplicate, which is a widely used procedure for evaluating the performance of a method in the absence of a reference material [38–40].

### 2.3. PAH Extraction

For the extraction procedure to recover the target analytes, an ultrasonic bath was used. This equipment is normally available in laboratories and has been used in different research works to extract PAHs from plant material [11,18,20].

An amount of 20 mL of a dichloromethane/hexane (1:1 $v/v$) mixture was added to each of the beakers (with the spiked and non-spiked sample) prepared in the previous step. The tops of the beakers were covered with aluminum foil and immersed in a 420-W ultrasonic bath for 10 min. This procedure was repeated two more times, using the fresh solvent mixture, for a total of 30 min of extraction and 60 mL of dichloromethane/hexane mixture for each sample. The three extracts of each sample were combined in a round bottom flask (100 mL) and evaporated on a Buchi rotary evaporator at 30 °C, with a pressure between 550 mbar and 170 mbar, to approximately 1 mL, and further cleaned-up.

### 2.4. Clean-Up and Final Concentration

Sep-Pak Alumina cartridges (6 cc, 1 g. Waters) were used for cleaning-up. Firstly, the cartridges were placed in a Waters SPE Vacuum Manifold and conditioned by passing 10 mL of the dichloromethane/hexane mixture through the bed with a flow rate of approximately 1.4 drops per second obtained by adjusting the vacuum. The mixture was collected in a test tube and was discarded. Then, the column was loaded with the extract and 10 mL more of the dichloromethane/hexane mixture was added to allow the elution of the analytes with the same flow rate of 1.4 drops per second, which were collected in a clean test tube. Finally, 5 mL of dichloromethane was added. The eluted extract was transferred to a round bottom flask (100 mL) to evaporate it again to approximately 1 mL under the same conditions indicated above. After that, the extract was transferred to 2 mL Eppendorf and concentrated to dryness using a Genevac miVac centrifugal concentrator at 40 °C for 30 min. Finally, the samples were reconstituted in 1 mL of acetonitrile, shaken, filtered using a PVDF syringe filter (32 mm, 0.22 µm) attached to a syringe of 3 mL, and transferred to 2 mL amber glass vial. This final filtration is carried out before the chemical analysis to avoid the obstruction of the HPLC column due to the presence of any particle.

## 2.5. PAH Analysis

The samples were analyzed by a HPLC (Agilent 1260 system) using a ZORBAX Eclipse PAH column (4.6 × 50 nm, 3.5 µm) and a UV detector (Agilent 1260 DAD G4212B) operating with wavelengths (λ) of 220 nm, 230 nm, and 254 nm. The column temperature was maintained at 25 °C, the injection volume was set to 20 µL, and the flow rate was 1.4 mL/min. The elution program was defined as follows (with acetonitrile (A) and water (B) as mobile phases): 0–6 min isocratic 40:60 ($v/v$) A:B; 6–9.5 min linear gradient from 40 to 100% of A and 9.5–12 min isocratic 40:60 ($v/v$) A:B. The peak intensity of each PAH changes depending on the UV wavelength; thus, the PAHs were calibrated at the wavelength where the intensity was greatest for that PAH. Table 1 shows the retention time and the UV wavelength at which the peak of each PAH was most intense. The PAH peaks in the sample chromatograms were identified by a retention time matching between standard and sample chromatograms. Quantification was performed by the peak area of each PAH using the ChemStation software (Agilent Technologies, Santa Clara, CA, USA).

**Table 1.** Retention time and the UV wavelength (λ) at which the peak of each PAH is most intense.

|  | Retention Time (min) | λ (nm) |
| --- | --- | --- |
| Naphthalene (Naph) | 3.0 | 220.0 |
| Acenaphthylene (Acy) | 3.4 | 230.0 |
| Acenaphthene (Ace) | 3.8 | 220.0 |
| Fluorene (FLuo) | 4.0 | 254.0 |
| Phenanthrene (Phen) | 4.3 | 254.0 |
| Anthracene (Ant) | 4.7 | 254.0 |
| Fluoranthene (Flt) | 5.0 | 230.0 |
| Pyrene (Pyr) | 5.2 | 230.0 |
| Benzo[a]anthracene (BaA) | 6.1 | 220.0 |
| Chrysene (Chry) | 6.3 | 254.0 |
| Benzo[b]fluoranthene (BbF) | 6.9 | 254.0 |
| Benzo[k]fluoranthene (BkF) | 7.3 | 230.0 |
| Benzo[a]pyrene (BaP) | 7.5 | 254.0 |
| Dibenzo[a,h]anthracene (DahA) | 8.2 | 220.0 |
| Benzo[g,h,i]perylene (BghiP) | 8.5 | 220.0 |
| Indeno [1,2,3-cd]pyrene (IcdP) | 9.1 | 230.0 |

## 2.6. Method Validation

Linearity, limit of detection (LOD), limit of quantification (LOQ), repeatability, and recovery were determined for validation of the HPLC method.

External standard calibration curves were obtained using the certified standard at eleven different levels in the concentration range of 2.5–2500 µg L$^{-1}$. The linearity of each PAH was evaluated as the coefficients of determination ($R^2$) by regression analysis.

Instrumental LOD and LOQ were calculated according to Equations (1) and (2), respectively [14,41,42]:

$$LOD = \frac{3.3\sigma}{IC} \quad (1)$$

$$LOQ = \frac{10\sigma}{IC} \qquad (2)$$

where IC is the calibration curve inclination and σ is the standard deviation of the intercept of the calibration curve with the *y*-axis.

Instrumental repeatability (precision) was studied as percent relative standard deviation (%RSD$_{inst}$) of three consecutive injections of the standard solution at 100 µg L$^{-1}$, while method repeatability was expressed as %RSD$_{method}$ of the concentrations determined in duplicate spiked samples.

The recovery values (%R) were determined according to Equation (3) [43]:

$$\%R = \frac{\text{PAH concentration in the spiked sample} - \text{PAH concentration in the non-spiked sample}}{\text{known added PAH concentration in spiked sample}} \times 100 \qquad (3)$$

## 3. Results and Discussion

### 3.1. Method Validation

The concentration range of calibration and linearity for each PAH, the instrumental limit of detection (LOD) and quantification (LOQ), instrumental and method repeatability expressed in terms of percent relative standard deviation (%RSD), and recovery values are reported in Tables 2 and 3. Moreover, as an example, Figures 1–3 show the chromatogram of the certified standard at 1000 µg L$^{-1}$, the spiked bark sample at 1.5 µg g$^{-1}$, and the non-spiked bark sample.

**Table 2.** Linear range, regression equation, coefficient of determination (R$^2$), instrumental repeatability expressed in terms of percent relative standard deviation (%RSD$_{inst}$), and instrumental limit of detection (LOD) and quantification (LOQ).

| PAH | Linearity | | | %RSD$_{inst}$ | LOD (µg L$^{-1}$) | LOQ (µg L$^{-1}$) |
|---|---|---|---|---|---|---|
| | Concentration Range of Calibration (µg L$^{-1}$) | Regression Equation [a] | R$^2$ | | | |
| Naphthalene (Naph) | 7.5–2500 | y = 2.4 ± 0.02x − 35.9 ± 21.53 | 0.9998 | 0.3 | 0.2 | 0.6 |
| Acenaphthylene (Acy) | 5.0–2500 | y = 1.1 ± 0.008x − 8.2 ± 9.0 | 0.9997 | 1.7 | 0.8 | 2.5 |
| Acenaphthene (Ace) | 10–2500 | y = 0.8 ± 0.008x − 6.8 ± 9.6 | 0.9996 | 0.2 | 2.6 | 7.8 |
| Fluorene (FLuo) | 10–2500 | y = 0.2 ± 0.004x − 3.7 ± 4.63 | 0.9993 | 0.006 | 5.5 | 16.8 |
| Phenanthrene (Phen) | 5.0–2500 | y = 0.9 ± 0.007x − 8.2 ± 7.14 | 0.9998 | 0.1 | 0.8 | 2.6 |
| Anthracene (Ant) | 2.5–2500 | y = 1.9 ± 0.02x + 1.3 ± 9.0 | 0.9996 | 3.8 | 0.7 | 2.2 |
| Fluoranthene (Flt) | 5.0–2500 | y = 0.6 ± 0.006x − 9.9 ± 6.4 | 0.9996 | 1.1 | 6.3 | 19.0 |
| Pyrene (Pyr) | 5.0–2500 | y = 0.7 ± 0.005x − 7.6 ± 5.6 | 0.9997 | 0.3 | 7.8 | 23.8 |
| Benzo[a]anthracene (BaA) | 10–2500 | y = 0.6 ± 0.006x + 0.4 ± 7.1 | 0.9996 | 1.0 | 1.6 | 4.9 |
| Chrysene (Chry) | 7.5–2500 | y = 1.2 ± 0.01x − 14.5 ± 11.1 | 0.9997 | 0.5 | 1.4 | 4.2 |
| Benzo[b]fluoranthene (BbF) | 7.5–2500 | y = 0.6 ± 0.006x − 5.3 ± 5.1 | 0.9996 | 4.6 | 4.8 | 14.7 |
| Benzo[k]fluoranthene (BkF) | 7.5–2500 | y = 0.5 ± 0.004x − 6.8 ± 5.2 | 0.9996 | 0.4 | 6.3 | 19.0 |
| Benzo[a]pyrene (BaP) | 7.5–2500 | y = 0.4 ± 0.004x − 1.4 ± 4.8 | 0.9995 | 4.3 | 1.2 | 3.7 |
| Dibenzo[a,h]anthracene (DahA) | 7.5–2500 | y = 0.6 ± 0.005x − 6.4 ± 5.8 | 0.9997 | 0.9 | 6.1 | 18.4 |
| Benzo[g,h,i]perylene (BghiP) | 25.0–2500 | y = 0.7 ± 0.008x − 14.4 ± 9.3 | 0.9997 | 1.2 | 6.5 | 19.6 |
| Indeno[1,2,3-cd]pyrene (IcdP) | 25.0–2500 | y = 0.4 ± 0.003x − 9.2 ± 3.8 | 0.9997 | 0.4 | 13.7 | 41.5 |

[a] Calibration curves constructed by linear regression of the peak area (y) of each PAH against their respective concentrations (x) (µg L$^{-1}$).

**Table 3.** Percentage recoveries (%R) and method repeatability expressed in terms of percent relative standard deviation (%RSD$_{method}$).

| PAH | Leaves | | Bark | |
|---|---|---|---|---|
| | %R | %RSD$_{method}$ | %R | %RSD$_{method}$ |
| Naphthalene (Naph) | 74.8 | 3.7 | 56.7 | 6.3 |
| Acenaphthylene (Acy) | 64.8 | 1.3 | 58.6 | 5.5 |
| Acenaphthene (Ace) | 67.9 | 3.9 | 55.2 | 12.8 |
| Fluorene (FLuo) | 75.1 | 1.9 | 72.1 | 13.8 |
| Phenanthrene (Phen) | 106.4 | 2.3 | 100.6 | 4.5 |

Table 3. Cont.

| PAH | Leaves | | Bark | |
|---|---|---|---|---|
| | %R | %RSD$_{method}$ | %R | %RSD$_{method}$ |
| Anthracene (Ant) | 88.9 | 1.7 | 92.4 | 3.1 |
| Fluoranthene (Flt) | 90.2 | 3.3 | 79.9 | 1.0 |
| Pyrene (Pyr) | 77.4 | 1.7 | 69.8 | 3.2 |
| Benzo[a]anthracene (BaA) | 85.1 | 1.2 | 86.7 | 2.9 |
| Chrysene (Chry) | 95.7 | 2.7 | 82.1 | 5.0 |
| Benzo[b]fluoranthene (BbF) | 72.1 | 3.1 | 75.7 | 1.6 |
| Benzo[k]fluoranthene (BkF) | 70.0 | 7.6 | 73.2 | 5.8 |
| Benzo[a]pyrene (BaP) | 91.2 | 16.9 | 69.3 | 11.0 |
| Dibenzo[a,h]anthracene (DahA) | 83.9 | 5.1 | 74.7 | 10.1 |
| Benzo[g,h,i]perylene (BghiP) | 76.4 | 24.0 | 82.8 | 10.5 |
| Indeno [1,2,3-cd]pyrene (IcdP) | 75.7 | 32.8 | 64.3 | 45.8 |

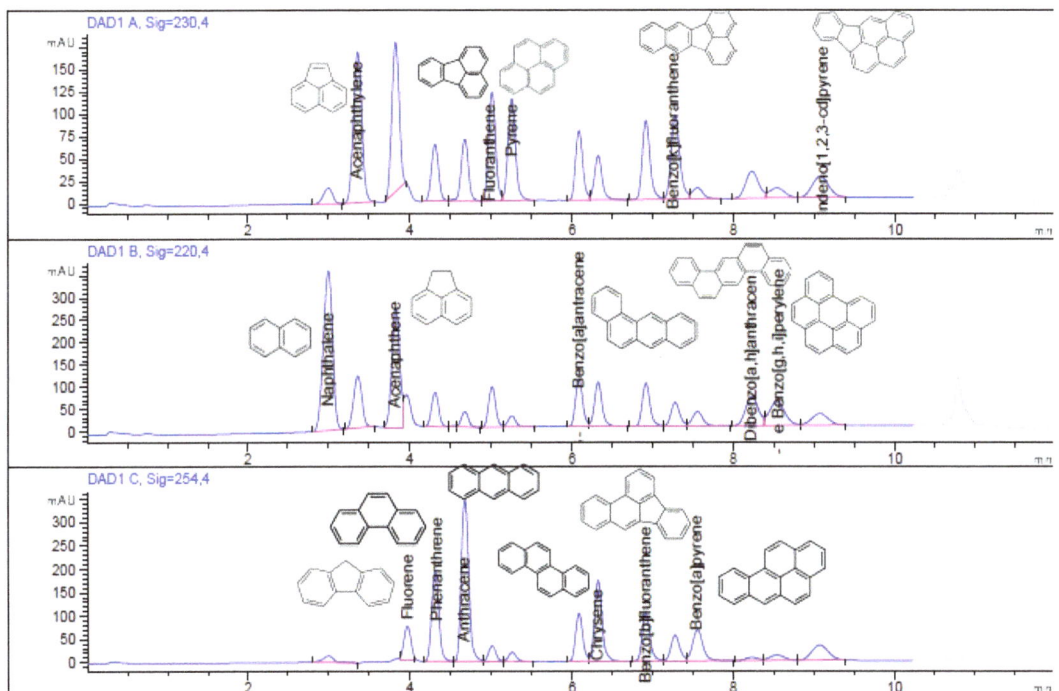

**Figure 1.** HPLC chromatogram of the certified standard at a concentration of 1000 µg L$^{-1}$ of 16 PAHs at λ = 230 nm, 220 nm, and 254 nm. Each PAH was calibrated in the UV wavelength (λ) where their signal was greatest.

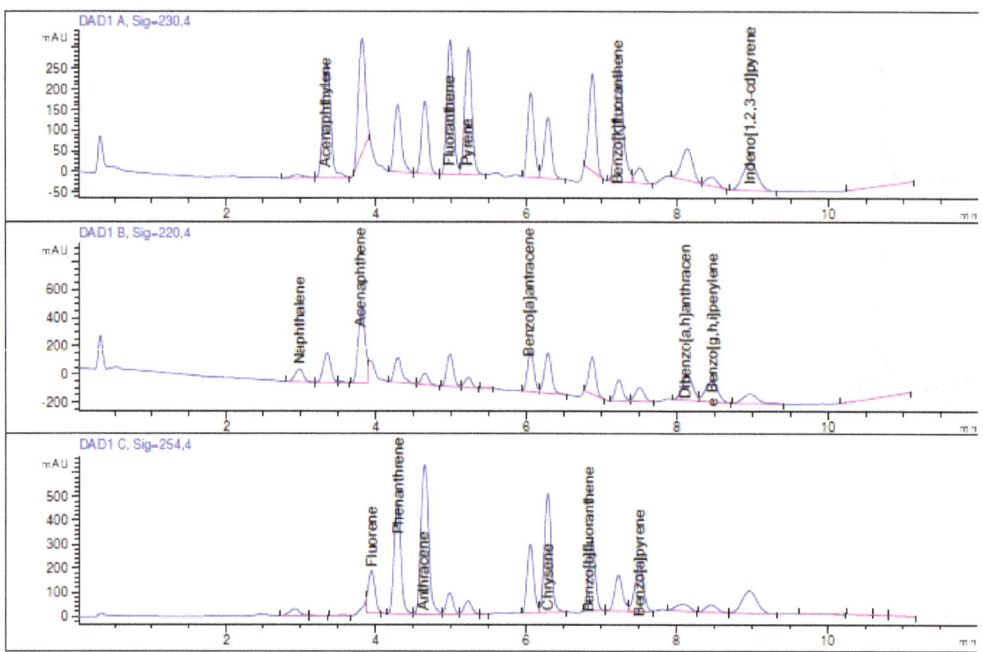

**Figure 2.** HPLC chromatogram of the spiked bark sample at 1.5 µg g$^{-1}$ at λ = 230 nm, 220 nm, and 254 nm.

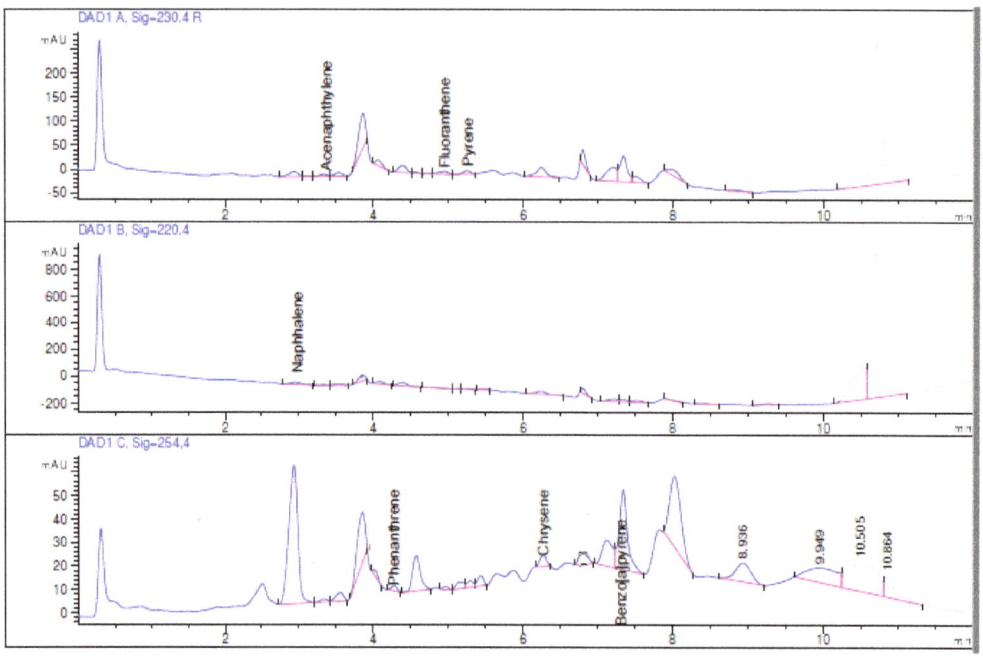

**Figure 3.** HPLC chromatogram of the non-spiked bark sample at λ = 230 nm, 220 nm, and 254 nm.

All calibration curves show good linearity with $R^2$ values ranging between 0.9993 and 0.9998 (Table 2). Instrumental LOD and LOQ ranged from 0.2 µg L$^{-1}$ for Naph to 13.7 µg L$^{-1}$ for IcdP, while LOQ values ranged from 0.6 µg L$^{-1}$ for Naph to 41.5 µg L$^{-1}$ for IcdP (Table 2). The instrumental repeatability values (%$RSD_{inst}$) ranged between 0.006% and 4.6% (see Table 2), being within the interval values found in the literature (1–47.3%) [44,45] and indicating good instrumental precision.

In general, the method repeatability (%$RSD_{method}$) was below 17% for the leaf and bark samples (Table 3), which is acceptable at such low concentration levels. An exception is observed for IcdP in both sample types and for BghiP in leaf samples. The high variability of high molecular weight PAHs has also been shown in previous works [46] and could be due to the remaining co-extracted interferences from leaves and bark. Moreover, the %$RSD_{method}$ values were similar to those reported in the literature (up to 18.8% [8] and 31.4% [44]).

Regarding recovery, the lowest values were found for the lighter PAHs, mainly Naph, Acy, and Ace (Table 3). The low recovery of the lighter PAHs could be due to the fact that they are more likely to be lost during sample handling and treatment, mainly in the evaporation/concentration step, and because lighter PAHs can penetrate further into the leaf tissues [47], making their extraction more complex. However, most %R values are within the 60–120% range, which is accepted as valid [48].

### 3.2. Real Contaminated Samples

Figure 4 shows the experimental data for the non-spiked leaf and bark samples, specifically: (a) the distribution of PAHs according to the molecular weight classification (light-molecular-weight PAHs (LMW: 2 and 3 rings PAHs), medium-molecular-height PAHs (MMW: 4 rings) and high-molecular-weight PAHs (HMW: 5 and 6 rings)); and (b) the individual PAH concentration. The concentration of each PAH was corrected based on the subtraction of the values of procedural blanks (extraction and clean-up of reagents without vegetative material) and the recoveries obtained in Table 3.

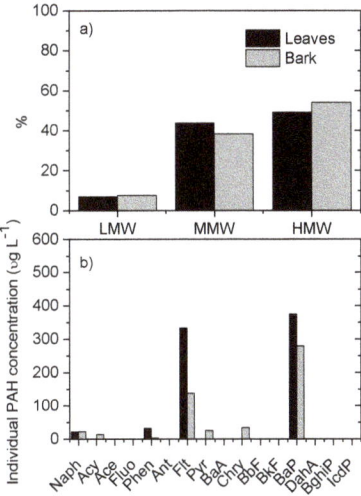

**Figure 4.** Experimental results for the non-spiked samples: (**a**) distribution of PAHs according to the molecular weight classification and (**b**) the individual PAH concentration.

It is observed in Figure 4a that there is a greater predominance of HMW PAHs, followed by MMW, and finally LMW PAHs. The high incidence of HMW PAHs is due to the high concentrations of BaP (Figure 4b), which could indicate a high human exposure risk, as this PAH is the usual marker of carcinogenic levels of PAHs in environmental studies [49]. BaP is mainly attributed to gasoline exhaust emissions that are known to contribute to more BaP emissions than diesel engines [50]. Moreover, in a recent work [51], it was found that BaP is associated with acceleration and braking activities, i.e., with the presence of speed-modifying devices, such as traffic lights, roundabouts, intersections, curves, and speed bumps. On the other hand, the high incidence of MMW is attributed to the high concentrations of Flt (Figure 4b), which is associated with coal combustion and biomass burning [52]. The incidence of LMW PAHs was not as high and could be related to their high vapor pressures (higher volatility) which causes them to be resuspended into the atmosphere [8,30].

## 4. Conclusions

An analytical method for the detection and quantification of the 16 EPA priority PAH in leaf and bark samples of *Sambucus nigra* was validated. The methodology combines ultrasonic extraction, subsequent concentration and clean-up, final concentration, and the chemical analysis by high-performance liquid chromatography (HPLC). Linearity of the calibration curve, instrumental LOD and LOQ, instrumental and method repeatability (precision) (%$RSD_{inst}$ and %$RSD_{method}$), and recovery experiments were used to validate the method. Good precision was observed, obtaining instrumental repeatability in the interval of 0.006–4.6%, while most of the method repeatability was below 17%, with exception of IcdP in both sample types (32.8% for leaf and 45.8% for bark) and BghiP in leaf samples (24.0%). Most recovery values were within the accepted range of 60–120%. However, lower values were obtained for Naph, Acy, and Ace, indicating that there has been a loss of these analytes during sample handling and treatment, which probably occurred during the concentration stage. Results from the actual contaminated samples indicated a high incidence in the air of fluoranthene, which is associated with coal combustion and biomass burning, and of BaP, which is a PAH highly associated with gasoline exhaust emissions and has a carcinogenic effect.

**Author Contributions:** Formal analysis, F.V., N.E.S. and K.A.; investigation, F.V., N.E.S. and K.A.; writing—review and editing, F.V., N.E.S. and K.A.; writing—original draft preparation, N.E.S. and K.A.; supervision, F.V. and K.A.; visualization, N.E.S. and K.A.; conceptualization, K.A.; methodology, F.V., N.E.S. and K.A.; validation, K.A.; resources, K.A.; data curation, K.A.; project administration, K.A.; funding acquisition, K.A. All authors have read and agreed to the published version of the manuscript.

**Funding:** This work was supported by Universidad de Las Américas (Grant number: AMB.KAF.21.02), Quito, Ecuador.

**Institutional Review Board Statement:** Not applicable.

**Informed Consent Statement:** Not applicable.

**Data Availability Statement:** Data are available from the corresponding author on request.

**Acknowledgments:** The authors express their gratitude to the Environment Ministry (MAE) for providing permission to collect the vegetable species samples (authorization number MAAE-ARSFC-2021-1255).

**Conflicts of Interest:** The authors declare no conflict of interest.

## References

1. Moret, S.; Conte, L.S. Polycyclic aromatic hydrocarbons in edible fats and oils: Occurrence and analytical methods. *J. Chromatogr. A* **2000**, *882*, 245–253. [CrossRef]
2. Moret, S.; Conte, L. A rapid method for polycyclic aromatic hydrocarbon determination in vegetable oils. *J. Sep. Sci.* **2002**, *25*, 96–100. [CrossRef]
3. World Health Organization–Office for Europe (EURO-OMS). *Air Quality Guidelines*, 2nd ed.; WHO Regional Office for Europe: Copenhagen, Denmark, 2000.
4. Dybing, E.; Schwarze, P.E.; Nafstad, P.; Victorin, K.; Penning, T.M. Polycyclic Aromatic Hydrocarbons in ambient air and cancer. *Air Pollut.Cancer* **2013**, *161*, 75.

5. U. S. Environmental Protection Agency Team. *Health Assessment Document for Diesel Engine Exhaust*; EPA/600/8-90/057F Report for National Center for Environmental Assessment Environmental Protection Agency (EPA): Washington, DC, USA, 2002.
6. Li, Q.; Li, Y.; Zhu, L.; Xing, B.; Chen, B. Dependence of plant uptake and diffusion of polycyclic aromatic hydrocarbons on the leaf surface morphology and micro-structures of cuticular waxes. *Sci. Rep.* **2017**, *7*, 46235. [CrossRef]
7. Fellet, G.; Pošćić, F.; Licen, S.; Marchiol, L.; Musetti, R.; Tolloi, A.; Barbieri, P.; Zerbi, G. PAHs accumulation on leaves of six evergreen urban shrubs: A field experiment. *Atmos. Pollut. Res.* **2016**, *7*, 915–924. [CrossRef]
8. Ratola, N.; Amigo, J.M.; Alves, A. Levels and Sources of PAHs in Selected Sites from Portugal: Biomonitoring with Pinus pinea and Pinus pinaster Needles. *Arch. Environ. Contam. Toxicol.* **2010**, *58*, 631–647. [CrossRef]
9. Huang, S.; Dai, C.; Zhou, Y.; Peng, H.; Yi, K.; Qin, P.; Luo, S.; Zhang, X. Comparisons of three plant species in accumulating polycyclic aromatic hydrocarbons (PAHs) from the atmosphere: A review. *Environ. Sci. Pollut. Res.* **2018**, *25*, 16548–16566. [CrossRef]
10. De Nicola, F.; Claudia, L.; MariaVittoria, P.; Giulia, M.; Anna, A. Biomonitoring of PAHs by using Quercus ilex leaves: Source diagnostic and toxicity assessment. *Atmos. Environ.* **2011**, *45*, 1428–1433. [CrossRef]
11. Ratola, N.; Lacorte, S.; Barceló, D.; Alves, A. Microwave-assisted extraction and ultrasonic extraction to determine polycyclic aromatic hydrocarbons in needles and bark of *Pinus pinaster* Ait. and *Pinus pinea* L. by GC-MS. *Talanta* **2009**, *77*, 1120–1128. [CrossRef]
12. Pereira, G.M.; da Silva, S.E.; Mota, E.Q.; Parra, Y.J.; Castro, P. Polycyclic aromatic hydrocarbons in tree barks, gaseous and particulate phase samples collected near an industrial complex in Sao Paulo (Brazil). *Chemosphere* **2019**, *237*, 124499. [CrossRef]
13. Niu, L.; Xu, C.; Zhou, Y.; Liu, W. Tree bark as a biomonitor for assessing the atmospheric pollution and associated human inhalation exposure risks of polycyclic aromatic hydrocarbons in rural China. *Environ. Pollut.* **2019**, *246*, 398–407. [CrossRef] [PubMed]
14. Ortega, L.M.; Uribe, D.M.; Grassi, M.T.; Garrett, R.; Sánchez, N.E. Determination of polycyclic aromatic hydrocarbons extracted from lichens by gas chromatography–mass spectrometry. *MethodsX* **2022**, *9*, 101836. [CrossRef] [PubMed]
15. Blasco, M.; Domeño, C.; López, P.; Nerín, C. Behaviour of different lichen species as biomonitors of air pollution by PAHs in natural ecosystems. *J. Environ. Monit.* **2011**, *13*, 2588–2596. [CrossRef]
16. Van der Wat, L.; Forbes, P.B.C. Comparison of extraction techniques for polycyclic aromatic hydrocarbons from lichen biomonitors. *Environ. Sci. Pollut. Res.* **2019**, *26*, 11179–11190. [CrossRef]
17. Protano, C.; Guidotti, M.; Owczarek, M.; Fantozzi, L.; Blasi, G.; Vitali, M. Polycyclic Aromatic Hydrocarbons and Metals in Transplanted Lichen (*Pseudovernia furfuracea*) at Sites Adjacent to a Solid-waste Landfill in Central Italy. *Arch. Environ. Contam. Toxicol.* **2014**, *66*, 471–481. [CrossRef]
18. Yang, M.; Tian, S.; Liu, Q.; Yang, Z.; Yang, Y.; Shao, P.; Liu, Y. Determination of 31 Polycyclic Aromatic Hydrocarbons in Plant Leaves Using Internal Standard Method with Ultrasonic Extraction–Gas Chromatography–Mass Spectrometry. *Toxics* **2022**, *10*, 634. [CrossRef] [PubMed]
19. Landis, M.S.; Studabaker, W.B.; Pancras, J.P.; Graney, J.R.; Puckett, K.; White, E.M.; Edgerton, E.S. Source apportionment of an epiphytic lichen biomonitor to elucidate the sources and spatial distribution of polycyclic aromatic hydrocarbons in the Athabasca Oil Sands Region, Alberta, Canada. *Sci. Total Environ.* **2019**, *654*, 1241–1257. [CrossRef]
20. Fasani, D.; Fermo, P.; Barroso, P.J.; Martín, J.; Santos, J.L.; Aparicio, I.; Alonso, E. Analytical Method for Biomonitoring of PAH Using Leaves of Bitter Orange Trees (*Citrus aurantium*): A Case Study in South Spain. *Water Air Soil Pollut.* **2016**, *227*, 360. [CrossRef]
21. Shukla, V.; Upreti, D.K. Polycyclic aromatic hydrocarbon (PAH) accumulation in lichen, Phaeophyscia hispidula of DehraDun City, Garhwal Himalayas. *Environ. Monit. Assess.* **2009**, *149*, 1–7. [CrossRef]
22. Lehndorff, E.; Schwark, L. Biomonitoring of air quality in the Cologne Conurbation using pine needles as a passive sampler—Part II: Polycyclic aromatic hydrocarbons (PAH). *Atmos. Environ.* **2004**, *38*, 3793–3808. [CrossRef]
23. Nascimbene, J.; Tretiach, M.; Corana, F.; Schiavo, F.L.; Kodnik, D.; Dainese, M.; Mannucci, B. Patterns of traffic polycyclic aromatic hydrocarbon pollution in mountain areas can be revealed by lichen biomonitoring: A case study in the Dolomites (Eastern Italian Alps). *Sci. Total Environ.* **2014**, *475*, 90–96. [CrossRef]
24. St-Amand, A.D.; Mayer, P.M.; Blais, J.M. Modeling PAH uptake by vegetation from the air using field measurements. *Atmos. Environ.* **2009**, *43*, 4283–4288. [CrossRef]
25. Ratola, N.; Herbert, P.; Alves, A. Microwave-assisted headspace solid-phase microextraction to quantify polycyclic aromatic hydrocarbons in pine trees. *Anal. Bioanal. Chem.* **2012**, *403*, 1761–1769. [CrossRef]
26. Navarro, P.; Cortazar, E.; Bartolomé, L.; Deusto, M.; Raposo, J.; Zuloaga, O.; Arana, G.; Etxebarria, N. Comparison of solid phase extraction, saponification and gel permeation chromatography for the clean-up of microwave-assisted biological extracts in the analysis of polycyclic aromatic hydrocarbons. *J. Chromatogr. A* **2006**, *1128*, 10–16. [CrossRef]
27. Schenck, F.J.; Lehotay, S.J.; Vega, V. Comparison of solid-phase extraction sorbents for cleanup in pesticide residue analysis of fresh fruits and vegetables. *J. Sep. Sci.* **2002**, *25*, 883–890. [CrossRef]
28. Navarro-Ortega, A.; Ratola, N.; Hildebrandt, A.; Alves, A.; Lacorte, S.; Barceló, D. Environmental distribution of PAHs in pine needles, soils, and sediments. *Environ. Sci. Pollut. Res.* **2012**, *19*, 677–688. [CrossRef]

29. Rodriguez, J.H.; Pignata, M.L.; Fangmeier, A.; Klumpp, A. Accumulation of polycyclic aromatic hydrocarbons and trace elements in the bioindicator plants Tillandsia capillaris and Lolium multiflorum exposed at PM10 monitoring stations in Stuttgart (Germany). *Chemosphere* **2010**, *80*, 208–215. [CrossRef]
30. Amigo, J.M.; Ratola, N.; Alves, A. Study of geographical trends of polycyclic aromatic hydrocarbons using pine needles. *Atmos. Environ.* **2011**, *45*, 5988–5996. [CrossRef]
31. van Drooge, B.L.; Garriga, G.; Grimalt, J.O. Polycyclic aromatic hydrocarbons in pine needles (*Pinus halepensis*) along a spatial gradient between a traffic intensive urban area (Barcelona) and a nearby natural park. *Atmos. Pollut. Res.* **2014**, *5*, 398–403. [CrossRef]
32. Rodriguez, J.; Wannaz, E.; Salazar, M.; Pignata, M.; Fangmeier, A.; Franzaring, J. Accumulation of polycyclic aromatic hydrocarbons and heavy metals in the tree foliage of Eucalyptus rostrata, Pinus radiata and Populus hybridus in the vicinity of a large aluminium smelter in Argentina. *Atmos. Environ.* **2012**, *55*, 35–42. [CrossRef]
33. Krauss, M.; Wilcke, W.; Martius, C.; Bandeira, A.G.; Garcia, M.V.; Amelung, W. Atmospheric versus biological sources of polycyclic aromatic hydrocarbons (PAHs) in a tropical rain forest environment. *Environ. Pollut.* **2005**, *135*, 143–154. [CrossRef] [PubMed]
34. Kołodziej, B.; Antonkiewicz, J.; Maksymiec, N.; Drożdżal, K. Effect of traffic pollution on chemical composition of raw elderberry (*Sambucus nigra* L.). *J. Elem.* **2012**, *17*, 67–78. [CrossRef]
35. Dawidowicz, A.L.; Wianowska, D.; Baraniak, B. The antioxidant properties of alcoholic extracts from *Sambucus nigra* L. (antioxidant properties of extracts). *Lwt* **2006**, *39*, 308–315. [CrossRef]
36. Schmitzer, V.; Veberic, R.; Slatnar, A.; Stampar, F. Elderberry (*Sambucus nigra* L.) Wine: A Product Rich in Health Promoting Compounds. *J. Agric. Food Chem.* **2010**, *58*, 10143–10146. [CrossRef] [PubMed]
37. Gentscheva, G.; Milkova-Tomova, I.; Buhalova, D.; Pehlivanov, I.; Stefanov, S.; Nikolova, K.; Andonova, V.; Panova, N.; Gavrailov, G.; Dikova, T.; et al. Incorporation of the Dry Blossom Flour of Sambucus nigra L. in the Production of Sponge Cakes. *Molecules* **2022**, *27*, 1124. [CrossRef] [PubMed]
38. Dugay, A.; Herrenknecht, C.; Czok, M.; Guyon, F.; Pages, N. New procedure for selective extraction of polycyclic aromatic hydrocarbons in plants for gas chromatographic-mass spectrometric analysis. *J. Chromatogr. A* **2002**, *958*, 1–7. [CrossRef] [PubMed]
39. Fernández, R.; Galarraga, F.; Benzo, Z.; Márquez, G.; Fernández, A.J.; Requiz, M.G.; Hernández, J. Lichens as biomonitors for the determination of polycyclic aromatic hydrocarbons (PAHs) in Caracas Valley, Venezuela. *Int. J. Environ. Anal. Chem.* **2014**, *91*, 230–240. [CrossRef]
40. Solgi, E.; Keramaty, M.; Solgi, M. Biomonitoring of airborne Cu, Pb, and Zn in an urban area employing a broad leaved and a conifer tree species. *J. Geochem. Explor.* **2020**, *208*, 106400. [CrossRef]
41. Fukuma. Resolution of the Collegiate Board–RDC N° 166. Available online: http://en.fukumaadvogados.com.br/wp-content/uploads/2017/08/RDC-166_2017-Analytical-Methods-Validation.pdf (accessed on 20 July 2022).
42. Lee, Y.-N.; Lee, S.; Kim, J.-S.; Patra, J.K.; Shin, H.-S. Chemical analysis techniques and investigation of polycyclic aromatic hydrocarbons in fruit, vegetables and meats and their products. *Food Chem.* **2019**, *277*, 156–161. [CrossRef]
43. Thermo Scientific. *Matrix Spiking—Why Spike and How to Do It*; Environmental & Process Instruments Division; Water Analysis Instruments: Illawong, Australia, 2011.
44. Ratola, N.; Alves, A.; Psillakis, E. Biomonitoring of Polycyclic Aromatic Hydrocarbons Contamination in the Island of Crete Using Pine Needles. *Water Air Soil Pollut.* **2011**, *215*, 189–203. [CrossRef]
45. Tham, Y.W.; Takeda, K.; Sakugawa, H. Polycyclic aromatic hydrocarbons (PAHs) associated with atmospheric particles in Higashi Hiroshima, Japan: Influence of meteorological conditions and seasonal variations. *Atmos. Res.* **2008**, *88*, 224–233. [CrossRef]
46. Ratola, N.; Lacorte, S.; Alves, A.; Barceló, D. Analysis of polycyclic aromatic hydrocarbons in pine needles by gas chromatography–mass spectrometry: Comparison of different extraction and clean-up procedures. *J. Chromatogr. A* **2006**, *1114*, 198–204. [CrossRef] [PubMed]
47. Bakker, M.I.; Koerselman, J.W.; Tolls, J.; Kolloffel, C. Localization of deposited polycyclic aromatic hydrocarbons in leaves of *Plantago*. *Environ. Toxicol. Chem.* **2001**, *20*, 1112–1116. [CrossRef] [PubMed]
48. Environmental Protection Agency (EPA). Compendium Methods for the Determination of Toxic Organic Compounds in Ambient Air: Compendium Method TO-11A. Available online: https://www3.epa.gov/ttnamti1/files/ambient/airtox/to-11ar.pdf (accessed on 4 February 2023).
49. IARC-International Agency for Research on Cancer. Outdoor Air Pollution/IARC Working Group on the Evaluation of Carcinogenic Risks to Humans. IARC-International Agency for Research on Cancer: Lyon, France, 2013. Available online: https://publications.iarc.fr/_publications/media/download/4317/b1f528f1fca20965a2b48a220f47447c1d94e6d1.pdf (accessed on 30 September 2019).
50. Wu, C.Y.; Cabrera-Rivera, O.; Dettling, J.; Asselmeier, D.; McGeen, D.; Ostrander, A.; Lax, J.; Mancilla, C.; Velalis, T.; Bates, J.; et al. An Assessment of Benzo (a) pyrene Air Emissions in the Great Lakes Region. *Arbor* **2007**, *1001*, 48104.

51. Alexandrino, K.; Sánchez, N.E.; Zalakeviciute, R.; Acuña, W.; Viteri, F. Polycyclic Aromatic Hydrocarbons in *Araucaria heterophylla* Needles in Urban Areas: Evaluation of Sources and Road Characteristics. *Plants* **2022**, *11*, 1948. [CrossRef]
52. Pulster, E.L.; Johnson, G.; Hollander, D.; McCluskey, J.; Harbison, R. Levels and Sources of Atmospheric Polycyclic Aromatic Hydrocarbons Surrounding an Oil Refinery in Curaçao. *J. Environ. Prot.* **2019**, *10*, 431–453. [CrossRef]

**Disclaimer/Publisher's Note:** The statements, opinions and data contained in all publications are solely those of the individual author(s) and contributor(s) and not of MDPI and/or the editor(s). MDPI and/or the editor(s) disclaim responsibility for any injury to people or property resulting from any ideas, methods, instructions or products referred to in the content.

MDPI
St. Alban-Anlage 66
4052 Basel
Switzerland
www.mdpi.com

*Methods and Protocols* Editorial Office
E-mail: mps@mdpi.com
www.mdpi.com/journal/mps

Disclaimer/Publisher's Note: The statements, opinions and data contained in all publications are solely those of the individual author(s) and contributor(s) and not of MDPI and/or the editor(s). MDPI and/or the editor(s) disclaim responsibility for any injury to people or property resulting from any ideas, methods, instructions or products referred to in the content.